for

CHARLES

Brilliant scholar, unswerving friend

BATTLES
OF
SOUTH AFRICA

TIM COUZENS

First published in 2004 in southern Africa by
David Philip Publishers, an imprint of New Africa Books (Pty) Ltd
99 Garfield Road, Claremont 7700, South Africa

ISBN 0 86486 621 6

Cover and text design: Alida Kannemeyer
Cover image: Painting by Charles D. Bell (MuseumAfrica)
Editor: Angela Briggs
Proofreader: Tessa Kennedy
Typeset in 10.5 pt on 13.5 pt Minion by Peter Stuckey
Printed and bound by Clyson Printers, 11th Avenue, Maitland

CONTENTS

LIST OF MAPS

INTRODUCTION

This book must start by declaring what it is not. It is not a complete compendium of South African battles, nor is it a perfectly weighted judgement on which were the most important.

It is not an in-depth expert military analysis of warfare in South Africa, nor is it particularly original in its methodology or much of the material it uncovers. It does not present an overall theory and, though there are some obvious linkages, it does not have a connecting central theme.

Its aims are much more modest. It is a personal selection of battles found to be in some way interesting. Some of them were critical, some curious. Some of them are almost completely unknown, certainly to popular consciousness. Who, for instance, has heard of Nakob or Amalinde, Centane Hill or Thaba-Moorosi? Others may ring a bell but their details are largely lost in the amnesia of history. What happened, when and where, at Boomplaats and Lattakoo, for instance?

With one or two exceptions the book does not cover the numerous battles of the Anglo-Zulu War or the Anglo-Boer War. There are any number of books on these campaigns, many of them easily accessible and they do a much better job than this particular author could ever hope to do.

Rather, this is a bedside book. Each chapter can be read like a short story. It is a labour of love and the author has visited all the sites involved. It is therefore a personal book of personal pilgrimages which has tried to preserve a personal touch. Those who do not approve of this should stop reading now. No other apology will be forthcoming.

The varying nature of the evidence has sometimes required different styles. The chapter on Khunwana, for instance, is a double, if not a triple, picaresque. That on Blouberg concentrates on the consciousness of a single participant. The story of Amalinde I found so astonishing I did not want to withhold my surprise. Thaba-Bosiu, on the other hand, is not a single battle but a series of events over a period of nearly a century in which the only consistent witness was the mountain itself.

Consistency of language usage also presents a problem. A choice has been made for what seems appropriate in each case. For example, different names for one and the same river have been used at different times and places – the Senqu, the Orange, the Gariep. Then again, certain archaisms have been preserved to present a picture of time past: the trial record of Slagtersnek, for instance, consistently spells names as Prinslo and Labuscagne.

One surprise might be that no photographs are included. There are several reasons for this.

Firstly, especially where battles are concerned, maps have a far greater fascination and allure. Secondly, photographs, paintings or drawings often present only a single, limiting image. It is far better to let the imagination play over the events, allowing time for reverie and for one's own feelings about how to interpret the events and how to insert oneself into them; seeing things from all sides and, perhaps, wondering how one would oneself have acted under the circumstances.

Thirdly, one of the major aims of this book is to encourage the curious few to visit the actual sites of these battles, to alert residents or visitors to interesting places that might be in the area or to stimulate others to travel specifically to places they may not have heard of or might not have considered going to. It might, also, make more interesting the places you pass without noticing them (such as Tafelkop). Qolora Mouth, for instance, might be pretty remote but Nongqawuse's pool is the stuff of myth and legend and the battle site of Centane Hill has the charm of benign neglect. Not far off the main road between Johannesburg and Cape Town, the battle of Boomplaats offers a diversion of a couple of hours (three at the most) and a chance to step aside from the rat race of the deadly highway.

An alternate route home (such as via Keimoes and Upington) can offer intellectual refreshment and the road less travelled is the one that usually provides the most excitement. Local travel should be accompanied by local reading. I have derived great pleasure from John Milton and Noel Mostert, Fred Cornell and Ian Bennett, André Brink, John Buchan and Sol Plaatje and urge everyone to prepare themselves in advance before visiting any of the battle sites. Deeper knowledge inspires deeper appreciation.

Standing on a battle site in those epiphanic moments when the sun is shining, or the wind is blowing, or the rain is raining, evokes a particular reaction – a consciousness of the contrast between the peacefulness of the present scene and the desperate and deadly struggle that once took place on that spot. Part of the pathos has to do with the folly of war. But that is perhaps to oversimplify. Perhaps conflict is part of the human condition which can never be eradicated. Past experience seems to indicate that this is so.

Some of the battles recounted here might perhaps have been avoided. Others changed the face of politics and of society. Some may seem to have been of no importance whatsoever.

Yet Darwin knew that even lowly worms can change the landscape and thereby influence, in however small a way, the course of history.

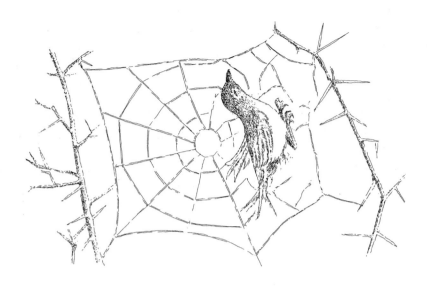

CHAPTER ONE

—•—

LOWER SABI

One should never be bored in the Kruger Park, no matter how often one goes.

There is invariably something beautiful to see (the blue flash of a lilac-breasted roller's wings in flight or the silent stealth of a crocodile drifting in a stream); something familiar (the sound of a hyena at night and guinea fowl at dusk); or something new (lions mating or honey badgers rutting, a giant African snail streaking across a road, or a cheetah's plaintive mewl).

Sometimes it is downright exciting. I have on occasion been charged by an elephant; once I was charged, at noon, by a leopard. (In Mkuze I have been chased up a tree by an irate black rhino.)

There is another curious phenomenon: the more one visits the park, more and more does it become peopled with memories. As one travels through it, or sits in silence, there is not only anticipation, there is also remembrance. Certain sights and events become fixed in the mind, their recall triggered by seeing a familiar tree, a river view, a prospect from a hill, a bend in the road. Here there was a leopard, there there was a watercolour sunset.

There have been many savage struggles to the death between animals, some witnessed, many more unseen in the bush or in the dark. Occasionally, as in the case of Ranger Wolhuter (who fought and killed a lion with a small knife) or anonymously with the human migration from Mozambique, animals have confronted man.

Occasionally, too, there is evidence of man taking on his fellow species in bitter contest. For instance, on the Crocodile River road (the S25) 14.5 kilometres south of the Biyamiti turn-off, there is an Anglo-Boer War monument recording the fact that on 16 September 1900 at that point south of the Crocodile River 'the Boer forces under General Ben Viljoen spiked and destroyed most of their artillery pieces and ammunition, to save them falling into the hands of the approaching British army under General Pole-Carew'.

A most curious, mysterious relic of early Portuguese penetration into the area is the rough Portuguese cross, the 'Das Neves' cross, carved into a leadwood tree on the S95 loop road north of Letaba.

Perhaps the most poignant battle I have personally witnessed in the park was on the Lower Sabi–Crocodile Bridge road – a small bird, caught in a spider's web strung between two bushes, struggling in vain with outstretched wings to free itself. I myself struggled with the impulse to intervene.

This microcosmic tussle I associate with a stone cairn carrying a bronze plaque because it was just a few metres away. The plaque, affixed to a stone cairn, is tantalisingly cryptic. 'On the 12th July 1725,' it reads, 'Frans de Kuiper and his 30 men were ambushed in this vicinity by Chief Dawano's warriors. The expedition was forced to retreat across the Lebombo range to the Dutch East India Company's fort at Lydzaamheid at Lourenço Marques.' Intriguing. When first I read it I wanted to learn more.

Little known (certainly to me) was the fact that the Dutch once occupied Delagoa Bay. When early into the eighteenth century the Portuguese at the port of Mozambique (in the north of present-day Mozambique) began to neglect Delagoa Bay (which was at that time only visited by dangerous pirates), the Dutch took the opportunity to step into the vacuum.

Rumours of gold in the interior were the main motivation. On 14 February 1720 two small ships, the *Kaap* and *Gouda*, accompanied by the *Zeelandia*, left Table Bay with a small band of soldiers, sailors and craftsmen under a clerk called Willem van Taak. They arrived at Delagoa on 29 March. The local Batonga chief, Maphumbo, subordinate to Mateke, welcomed them cordially.

The Dutch set about constructing a pentagonal earthwork fort which was named Fort Lagoa. Within weeks two-thirds of the over 100 colonists were dead of fever. In August the *Zeelandia* brought 80 soldiers as reinforcements.

A trickle of copper, iron, gold and ivory coming in as trade goods prompted enough interest at the station to equip an expedition of 19 men under Jan Stefler to venture into the interior in August 1723. They reached pleasant countryside in the Lebombo mountains before being attacked by local tribesmen who forced them to turn back.

In May 1724 Jan van de Capelle, a junior officer, who proved more resilient than most against fever, was appointed *secunde* (deputy-governor) at Fort Lagoa in Delagoa Bay. Persistent hints of a rich gold-field – associated with the great empire of Monomotapa – led him to send another party to search for it. Sergeant Johannes Monna led 31 men including a clerk who was assigned to keep a journal. His name was Frans de Kuiper (François de Cuiper). (Some of the places he subsequently recorded are identifiable; others are lost in the mists of history.) They set out on Wednesday, 27 June 1725. Aside from Monna and De Kuiper there was a corporal, a drummer, an under-master, an overseer of the pack-animals, twenty soldiers and five sailors. There were also eight interpreters and servants. For the transport of the provisions there were ten pack-animals; for dinner on the hoof there were eight head of cattle.

At first they passed through the territory of Chief Mattolle (Mathole) and the locals took to their heels at the sight of them. It was not an auspicious beginning. When they came across a river, probably the Matola, they named it the 'Oliphants' (because they saw plentiful signs of the eponymous animal), and spent a day exploring it. An encounter with locals in the area of Muambo (in the territory of Chief Semane) led to the chief promising to supply guides to take the Dutch to Nassangano, who seems to have been their immediate goal.

When they reached the Komati River on 1 July one of their party went to visit Nassangano's brother-in-law whose kraal was nearby and returned with the ominous report that the chiefs had instructed their people to murder the travellers and punish their guides. The brother-in-law was brought in to the camp and he confirmed the report. He also said that there was some gold and a lot of silver at an area he called 'Coupane Ciremandelle', much more gold at 'Beloele' and gold, copper and ivory at Simengele. Coupane was, he added, six days away, two of which were without water. He did not want to lead the expedition there.

The Dutch rested the next day, invited Semane to the camp and offered him a drink and snuff. The first he disliked; the second he enjoyed enormously. Half-a-thousand of his followers looked on with curiosity, many seeing white people for the first time. So the drummer was asked to play, and the chief called up more drums from the village, and the dancing went on until sunset.

The expedition travelled past Mkangane and Makanje and crossed the Sonduene River (perhaps the Masondwini or Klip). On Thursday, 5 July, they sent ahead scouts who were told that Coupane was 'three hours away' but over very difficult mountainous terrain and a river crossing which was too deep to manage. In fact, the doubting Dutch found another crossing, only four feet deep and a pistol-shot wide, further upstream. (One wonders if this was not the Crocodile.) They had come to realise that the local population 'were very unhappy with us having come there to explore the land'.

Over the next few days Chief Coupane proved elusive. They were told he had left for the mountains (the Lebombo?) to buy cattle, having instructed those left behind not to provide any assistance to the newcomers. They were told he would return 'in three days'. When they proceeded they heard that he was in the nearby settlement of his father, Alari Motsari. Then they heard that he had left that morning but would join them at the Monganje River. He never did.

After they crossed the Monganje on the 10th, the air of menace increased. The journey was now accompanied by continuous shouting, blowing of horns and rattling of assegais against shields. The party prepared itself for the worst. When they enquired about the surrounding restlessness they were told they must give up their guide and find their own way. They did, however, acquire a replacement and that evening reached a stream close to the settlement of Dawano.

Next day they were forced to stay put owing to the exhaustion of the pack-animals. They asked the headman about the road to Ciremandelle, approaching the matter as diplomatically as possible. But as often as they asked so often did they receive a different answer: five days' journey with no water for two days; seven days with some of them without water; a few days with water in abundance! The Dutch wanted to buy cattle; two animals were brought but when the sellers saw the barter beads they spat and left with the animals, saying they did not want these and preferred small blue ones, big yellow ones or large white ones.

Indeed, De Kuiper had found the same thing ever since they had entered the land of Nassangano – they had been unable to buy a single animal for the beads they had. Even at Coupane's five big bunches had been demanded for a single animal …

That night there was thunder, and lightning, and rain.

Next morning (Thursday the 12th) they still could not proceed because of the condition of the animals so De Kuiper questioned an elderly man, who claimed to have travelled extensively beyond the land of Thowelle. De Kuiper softened him up with a piece of meat and a drink. He was told that it was one day's journey to the Sabi River, another day to the kraal of Massawane, another to Matome's, two more to Ciremandelle – five days in all. Ciremandelle was the place of copper. You had then to cross ten rivers and pass many settlements before you reached the place of the Great Chief Intowelle where some gold was produced. But he highlighted Tsouke as the most important spot of all, where Portuguese ships would arrive to do business. He had himself seen the chief of Tsouke in his golden armour, and all his subjects armed in copper. At Bombane the inhabitants even wore gold necklaces. After an hour the old man grew bored and left with a small gift.

Now the Dutch realised that further progress was futile. Their animals were weak; their beads were useless for barter; the locals were increasingly hostile. Even more serious was the fact that one of the pack-animals had fallen into one of the rivers and two-thirds of their gunpowder was spoilt.

As De Kuiper sat bringing his journal up to date at about four o'clock that afternoon, with his gun resting in front of him, he suddenly heard an alarmed cry: 'Look out! Look out! Shoot!'

He jumped to his feet, grabbed his gun and passed the flintlock to the sentry who began firing, straightaway killing a warrior who was storming at them with shield and assegai. At the same time they could see attackers round them as numerous as the grasses of the veld. They were making such an awful noise with their whistling and blowing of horns and screams that the defenders could not understand what they shouted to each other, no matter how close they were. But they closed ranks and fired bravely at the 'dogs', who wanted to tear them apart like angry lions. With their first volley they killed six of Dawano's men and badly wounded at least ten others. The attackers continued to hurl their assegais with great courage.

Two of the Dutch were seriously wounded. They had been ambushed in a thicket close to the encampment by members of Dawano's vanguard who had stalked them and wanted to steal their guns.

The attack had evidently been planned with care. While the main body diverted the Dutch, others made for the cattle, though when these stampeded towards the camp they were killed with assegais. Seeing this, the Dutch retreated to open ground. The attackers surrounded them but, seeing the devastation caused by the gunfire, kept their distance while maintaining all the while their terrifying cacophony. Because of the serious shortage of gunpowder Monna ordered his men to desist from shooting unless a point-blank target presented itself. The Dutch began an orderly retreat to the mountains. They were followed until about seven that evening, when it was completely dark, assailed all the way by spears and stones.

When they were sure the 'bloodhounds' had abandoned the chase, the shaken party decided to move into the mountains, marching in the cold night over terribly big boulders and through rivulets until two in the morning when they broke through into flat country and could hear the barking of dogs from kraals in several directions. In order not to attract attention they lay still. This they did for over an hour until they were so dismally cold that their teeth began to rattle, and they all began to shake. So they got up and marched with their God's guidance, and were led by the stars through rocks, and bushes, and fields.

They did not encounter anyone until eight that morning when they came across a large settlement with its inhabitants drawn up in front in a hostile manner. The Dutch, prepared to shoot, asked for directions to the Komati River, which was to the 'right' (south). They marched steadfastly on till midday when they encountered five or six big kraals where they were welcomed once more with assegais and stones. De Kuiper asked them, through an interpreter, the reason for their aggression. Because the Dutch were passing through their land, they replied, and because they wanted everything that belonged to the Dutch including their flintlocks, which they called 'hongo songilo' – 'beautiful sticks'. The sergeant threw two bunches of beads towards them, and the soldiers threw some other trinkets thinking that this might make them desist from their threats. Instead, they just shouted 'like mad dogs': 'Hotte hotte molonge' ('Let us devour the white people'). When the Dutch heard this they took aim and shot seven of their assailants, killing six and wounding one in the leg.

At this the warrior band took fright – there were about 500 of them – and most fled. Now they were afraid of the sticks they had coveted. Only a few followed in the bush at a safe distance and the Dutch reached the Komati River within half-an-hour. The river was about half a cannon shot ('*carbijn schoot*') wide at that point and as the party waded across a number of their pursuers shouted over to the other bank warning the inhabitants of the lethal nature of these intruders. The Dutch consequently marched on unhindered and after 20 Dutch miles they arrived at dusk at a great valley where a man from Matekij greeted them and extended them much kindness.

It was Friday the 13th.

Next day, when they were again threatened, they told their interpreters to warn their shadowers that, at the slightest sign of attack, they would kill them all and burn down their houses ('*caasjes*'). They had no further trouble.

At daybreak on the 16th, they arrived at the settlement of Matolle and a little after at Maphumbo, happy that they were nearing home. Here the inhabitants greeted them with great joy, the women tearing branches from the trees and waving them above their heads. Others used the branches to sweep clean the path before them. Van de Capelle had sent his sergeant-major to accompany them the rest of the way and he had brought bread and some arak – that coarse but potent liquor from the East – which lifted their spirits enormously.

They marched straight to the fortress in Delagoa Bay, where they were received by the commander. They had to report that they had done their duty but that their expedition had failed.

Other expeditions by land and river also returned empty-handed. The fort turned out to be too small for the purposes asked of it and a larger one called Fort Lydzaamheid was built. But the tiny mosquito was proving the most deadly enemy of all, causing a terrible death-rate, and some investigation was made into whether Inhambane would not be a healthier spot for settlement.

So serious was the misery that a large section of the garrison at Fort Lydzaamheid set up a plot to seize the magazine, arm themselves, dispatch any opposition, and head for the Portuguese factories to the north. The *secunde* Van de Capelle got wind of the scheme via information given him by a 'sailor boy' and 62 men – a third of the garrison – were placed under arrest.

They were brought before the special court and were prosecuted by Ensign Monna (the same who led the De Kuiper expedition?). There was considerable apprehension related to the trial since the loyalty of the guards was almost as suspect as that of the prisoners. Over half the accused were sentenced to death but lots were drawn allowing commutation for some to a future of hard labour in chains.

In the end 22 of the mutineers were executed. Some were fixed to crosses, their bones were broken with iron bars, then their heads were cut off; some were suffocated close to death and then beheaded; others (the lucky ones?) were simply hanged.

Close on the heels of the trial came another calamity. A coalition of chiefs – Matashaj, Mambe, Matoli and Mateke – attacked first Maphumbo, then Chief Kwambe, close to a small Dutch outpost. Van de Capelle immediately responded and on 30 April 1729 sent a party of 29 soldiers and a few slaves under Ensign Monna to punish the incursion. Rashly the scouts under Sergeant Jan Mulder fired on the coalition warriors and the whole Dutch party was wiped out, except for a single slave survivor.

The cost of the whole enterprise and the failure to produce tangible results eventually caused the council of the Dutch East India Company to cut its losses. On 27 December 1730 the entire population of the settlement, including Van de Capelle, was taken on board the ships *Snuffelaar*, *Zeepost* and *Feyenoord* and repatriated to Cape Town. The loss of life over the decade of occupation had been appalling.

The skirmish with Dawano was small in scale, but had not the check to Monna and De Kuiper occurred the outcome of the Dutch occupation at Delagoa Bay might have been very different. Indeed, if one thinks about it, a permanent Dutch presence there might have ultimately given very different support to the future Boer republic in southern Africa.

So that plaque between Lower Sabi and Crocodile Bridge camps is worth a stop. As you sit there let your imagination re-enact that short, sharp engagement between Dawano and the Dutch.

CHAPTER TWO

———•◆•———

SALDANHA BAY (1)

Conflict was not unknown to Saldanha Bay in the early years. Very early into their rule – after 1652 – the Dutch joined the lucrative practice, initiated by the French, of slaughtering seals by the thousands in the Bay. During a minor mutiny on board a French ship, the Dutch galiot *Roode Vos*, was sent to investigate and, in the course of the expedition, the cook of the galiot (one Buijsman) was stoned to death on the shore by some local Khoi.

In the mid-1660s the French, pursuing the ambitions of Colbert, financial advisor to King Louis XIV, seriously considered formally annexing the Bay in order to establish a replenishing station for its East India Company fleets. After an initial sortie by the Marquis de Mondevergue in the *St Jean* in 1666, Admiral de la Haye was sent in 1670 to survey the Bay and examine the possibility of establishing a base. Six of his ships put in at Saldanha in August and humiliated the sergeant, Hieronymus Cruse, in charge of the handful of Dutch troops stationed there. They confiscated his belt and forced him to hold up his trousers while answering questions. Cruse did manage to escape and made his way (no doubt with his trousers secured) to Cape Town to warn his superiors there. They made ready to resist De la Haye but the admiral, having put up a marker proclaiming the Bay as French territory, headed for Madagascar instead.

Fighting with local groups also occurred. In 1672 followers of the Cochaqua chief Gonnema, after ambushing a hunting party at Twenty Four Rivers, attacked some unsuspecting Dutch on the shore of the Bay and beat them to death. This war only petered out in 1677. Fighting between local groups was also not unknown. In 1689 a force of Charingurikwa and Namakwa pounced on the Cochaqua settlement at the Bay and devastated it and its population.

But Saldanha has also been the site of two of the biggest naval battles in South African history.

In December 1780 the Dutch had been drawn in on the side of the French and the Americans against the British. Immediately the Cape became an important strategic target for both sides. The British prepared a fleet to sail as soon as possible. They took advantage of a privateering enterprise that was about to set out for the River Plate (in South America) under the command of Commodore George Johnstone.

They may have been the first at the starter's flag but they suffered from two significant handicaps.

Firstly, the Commodore was not only of relatively lowly rank but he was inexperienced and less than talented. Secondly, a spy in London called De la Motte fed the French with full details of the existence, composition and destination of the British fleet (he was later detected and executed).

Johnstone's privateering group was considerably strengthened and when it set sail from Spithead on 13 March 1781 it comprised 46 ships. There were 5 ships-of-the-line: the *Hero* (74 guns), *Monmouth* (64), *Jupiter* (50), *Isis* (50) and the flagship *Romney* (50); there were four frigates: *Apollo* (38), *Jason* (36), *Active* (32) and *Diana* (28); there was a fireship (the *Infernal*) and a bombship (the *Terror*) and a variety of smaller warships, four transports and 13 Indiamen. Three thousand soldiers, under the command of General Meadows, accompanied them.

Johnstone first made for St Jago (São Tiago) in the Cape Verde islands in order to take on fresh water and supplies. He anchored in the Porto Praia roads, but his inexperience (coupled with his naive belief that there was no danger) led him to dispose his fleet as ineptly as the Americans did at Pearl Harbour – his warships were stationed inside the convoy and the convoy scattered in the outside lines. Experienced seamen were sent ashore and the decks of the fighting ships were laden with casks and other impedimenta. Johnstone even ignored the implicit warning in the Port book at Praia mentioning the visit of a French frigate some weeks before which had warned the inhabitants that a French fleet would visit in April and that they should have cattle and other supplies ready for it. Yet Johnstone saw no danger.

On 22 March Acting Commodore Pierre André de Suffren, a determined and able man, sailed from Brest. His five ships-of-the-line, the *Heros* (74 guns), *Hannibal* (also 74), *Artésien*, *Vengeur* and *Sphinx* (all of 64 guns), actually outmatched the British, though Suffren was without a screen of fast and flexible frigates. He, too, had transports loaded with troops.

On 16 April the unsuspecting British at Porto Praia observed ships flying no identifying colours approaching. The *Heros*, *Hannibal* and *Artésien* anchored close to the *Isis* and their broadside into the frigate signalled their adversarial intent. They then ran up the French colours. The action that followed was a somewhat messy affair. It took the British some time to recover from their surprise but after the initial broadside on the *Isis* they began to fight back. Captain Ward of the *Hero* led a boarding party on the *Artésien*, killed its captain, and captured 25 of the crew but not the ship itself.

Cannon broadside after broadside lasted for two hours with neither side significantly getting the upper hand until the loss of its mizzen, main and fore topmasts threatened the loss of the *Hannibal*. So Suffren cut his cables and stood out to sea, with the *Heros* towing the crippled *Hannibal*. The French took with them as prizes two Indiamen, the fireship and a victualler (they were all recovered within days since they were in such bad shape that the French abandoned them).

The British prepared to pursue the French but it took them several hours to get under way and by then their foe had gone. The action was indecisive – but the race was on for the Cape. Johnstone surmised that Suffren would send part of his damaged fleet to the West Indies or make for Brazil to repair them and resupply. Again he was mistaken. The competent French commander did not flinch from his goal: he temporarily remasted the *Hannibal* and headed south.

Meanwhile, at the Cape itself, the governor, Joachim van Plettenberg, found himself in an unenviable position. Both fleets had actually left their home bases before the French frigate *Silphide* brought the news to Table Bay (on 31 March) that the Dutch were at war! Van Plettenberg's resources were slim – he had at his disposal scarcely more than 500 soldiers, some of whom were working on the scattered farms, and a few hundred assorted civil servants, mechanics and recuperating sailors. The garrison commander was the recently arrived Captain Robert Jacob Gordon (who had the previous year explored to the northern edge of the Colony, discovered a river there, raised a Dutch flag over it and called it the Orange) and second-in-command was Captain Carel Mattheys Willem de Lille, who had been in the colony for ten years. The governor, in theory, could also draw on the military services of 3 000 burghers. But not all of these, or even very many of them, were readily available. The perennial small-scale attacks of Bushmen meant many farmers were reluctant to leave their farms and families. In 1772, for instance, a farmer and his wife, daughter and a servant were murdered by a party of Bushmen in the Roggeveld. A commando tracked them down to a mountain cave and stormed it, killing 6 Bushmen and capturing 58. These were taken to Cape Town where the men were tried; one was broken on a wheel, one hanged and four were flogged, had the tendons of their heels severed and given hard labour for life. The women and children were given out to various individuals as servants. Bushmen were hunted down by several commandos and over 500 were shot and over 200 captured. But the threat was continuous and many individual confrontations occurred which were never recorded – one-on-one shootouts which ingrained themselves in the psyche of the farmers as deeply as they did in the American Wild West.

Furthermore, from 1779, strife with the Xhosa on the eastern frontier tied down considerable manpower.

So Van Plettenberg's chances of resisting a strong force of well-drilled British regulars were not good. His only real hope lay with the French. The wait for news was nail-biting. He spent his time trying to protect or hide the ships which found themselves in the vicinity of the Cape.

The approach of winter rendered Table Bay unsuitable and Simon's Bay could not be defended from an English fleet. So Van Plettenberg built a battery on the western point of Hout Bay and mounted 20 cannons there. There, too, he sent the *Batavia*, *Amsterdam*, *Morgenster* and *Indiaan*, all on their way home laden with rich cargoes from the east. But Hout Bay, relatively safe haven as it was, was too small to accommodate more large ships.

So, on 13 May, a homeward-bound convoy consisting of the ships *Hoogkarspel*, *Honkoop*, *Middelburg*, *Dankbaarheid* and *Paarl* was sent to Saldanha Bay, together with a Ceylon-bound Indiaman, the *Held Woltemaade*, which needed urgent repairs. Captain Gerrit Harmeyer, of the *Hoogkarspel*, was in command of the squadron. He was under instructions to destroy his ships if they were in danger of falling into enemy hands. The ropes and sails of the ships were sent further up the lagoon in the packets *Zon* and *Snelheid* near to Schapen Island. They were to be burnt if necessary in order to render the large ships immobile and therefore useless if threatened. The French traveller and botanist François le Vaillant, who was in Cape Town at the time, persuaded Captain van Gennep of the *Middelburg* to take him on board. The kings of Ternate and Tedore were prisoners on board one of the other ships.

At last Van Plettenberg's anxiety was partly alleviated on 20 May when the French frigate *Serapis* glided into Simon's Bay with the welcome news that the arrival of Suffren's fleet was imminent. This was correct. Suffren's hare had beaten Johnstone's tardy tortoise. The *Heros* sailed into Simon's Bay in advance of the rest of the French. A few days later transports brought in the Pondicherry infantry regiment and an artillery company to strengthen, at least temporarily, the land defences. They marched to Cape Town, which they entered on 3 July.

Ships sailing north were stopped to prevent information reaching the English but the *Held Woltemaade* was allowed to leave Saldanha Bay on its outward journey to Ceylon. This was to prove a fateful decision for the convoy holed up in the Bay.

On the morning of 22 July Van Plettenburg got a report from Saldanha that a large fleet had been spotted outside the Bay's entrance and he sent Lieutenant van Reenen, a burgher, with a troop of mounted men to reconnoitre the scene, while the ships at Hout Bay were ordered to Table Bay with instructions to stay close to shore and beach themselves in an emergency.

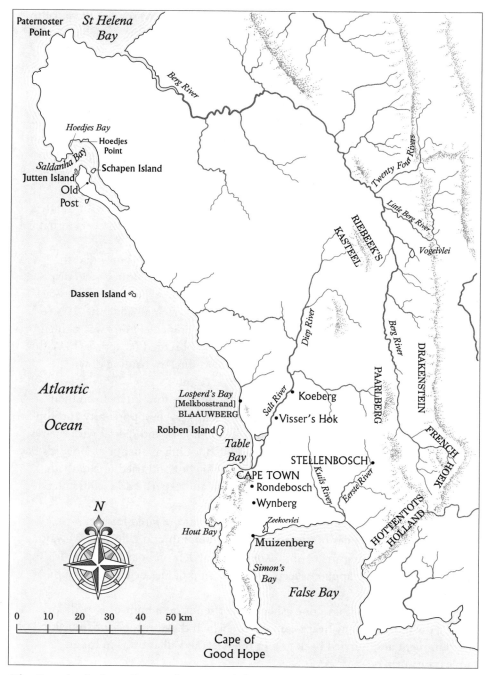

The Cape in the late nineteenth century (after G.M. Theal)

Johnstone meanwhile dispatched the *Active* to Table Bay to spy out the lie of the land – and sea. Hardly had the *Active* forged its way ahead (flying French colours) than it flushed an unknown sail heading southward. Deceived by the *Active*'s disguise, the *Held Woltemaade* submitted without a fight and gave Captain McKenzie vital intelligence about the ships and their dispositions within Saldanha Bay.

On 8 July Johnstone had suggested that he might put in at Saldanha Bay instead of Table Bay, determine whether the French had arrived in the Cape, and decide whether an overland advance on Cape Town might be possible. General Meadows nixed this in no uncertain terms: 'I have no objection to going to Saldanha but the greatest to landing with so small a force 40 miles from the Cape [something of an underestimate], in enemy territory, without water, through difficult passes and with no possible convoys or contact with the fleet. If I commanded at the Cape with the country on my side it is the very situation I should wish for my enemy.'

Now, with the capture of the *Held Woltemaade*, the British knew that the garrison at the Cape was too strong for the size of the force at their own disposal. So, with Lieutenant William Patterson – a man who had experience of the west coast – as his pilot, Johnstone decided to attack the Dutch at Saldanha Bay. At nine-thirty on the morning of 21 July a fleet flying French colours was sighted by lookouts at Saldanha Bay, close in because fog had obscured its approach. Within an hour the lead ships were inside the entrance to the Bay, hauling down the French flags and replacing them with their true colours.

For the British, speed and surprise were essential as they knew from their captured prize that the Dutch would fire their ships given opportunity and time. The *Romney* was in the lead, followed by the *Jason*, *Lark* and *Jupiter* with the rest of the squadron bringing up the rear. The Dutch, riding at anchor at Hoedje's Bay, panicked, hastily set their ships alight, weighed anchor and tried to run them aground. But, by then, the British were on them and expertly dowsed the fires while the Dutch crews swam for shore.

Only Captain van Gennep of the *Middelburg* was prepared for every eventuality. Large bundles of tow soaked in tar, sulphur and tallow were awaiting the torch which the first mate, Abraham de Smidt, left on board precisely for this purpose, immediately applied. Deep inside the ship flames began to consume it from within.

Le Vaillant had that day gone ashore with friends on a hunting expedition. During the morning they heard cannon-fire which they mistook for some kind of entertainment and hurried back to join the fleet. Le Vaillant was in for an unpleasant surprise.

Having secured four of the five ships the British turned all their attention to the stricken *Middelburg*. Johnstone himself was in the boarding party approaching the ship, which was by now almost engulfed by 'streaks of fire that broke from the main body of the flame loudly snapping like the reports of bombs, encircled stem and stern, cracking from the main-tops, bidding defiance to the most intrepid'.

That they did get close was lucky for two English prisoners, chained below deck, whose plight had been overlooked by the retreating Dutch. Their cries for help were heard and, in a magnificent act of selfless and unrewarded heroism, an unknown sailor jumped on board with a crowbar and wrenched the ringbolt from its mooring. The prisoners escaped through a porthole but their rescuer was badly burnt and lost the use of an arm.

The burning *Middelburg*, freed of its moorings, was now a mortal threat to the other prizes so Johnstone closed with the blazing hulk and secured a line to it. He was joined by many other boats and soon, like a tug-o'-war team matching themselves against a great deadweight, they were towing the *Middelburg* away towards Hoedje's Point, where they abandoned it. Just in time. Ten minutes later the flames reached the magazine and a huge explosion ripped it apart.

This event was a nightmare for Le Vaillant, who had just arrived on the scene. He watched with horror: 'The *Middelburg* blew up and in a moment the sea and sky were filled with burning papers. I had thus the cruel mortification of seeing my collections, my fortune, my projects and all my hopes rise to the middle regions and evaporate into smoke.' One can hardly imagine a worse experience for an ardent scholar than to see his years of research disappear in an instant in front of his eyes.

Outside the main action two secondary initiatives occurred. The sloop *Rattlesnake* raced up the bay making for Schapen Island and the two packets on which the vital sails of the Indiamen were stowed. The *Zon* and *Snelheid* were captured without any attempt on the part of their crew to scuttle them. And landings were made at Riet Bay (in order to occupy the by now burning Company post) and Baviaans Bay (or North Bay) at the entrance to Saldanha Bay. These landings did not materially affect the course of the battle but by two o'clock in the afternoon a company of Grenadiers under Major Hutchinson occupied Hoedje's Point and captured the signal guns. Broadsides from the British warships pursued the Dutch refugees in their flight to Cape Town. The royal oriental prisoners were released and treated with honour.

As so often in such cases the British victors had triumphed in the battle but their main purpose had been thwarted. Johnstone split up his force – a convoy with the troops and escorted by five warships was sent on to India. The rest of the main battle fleet he took with him back to England. Ironically, the five prizes from which he and his men stood to benefit as spoils of war he dispatched to St Helena. A hurricane sank three of them and only the *Hoogkarspel* and the *Paarl* made it to England.

Today the whole Saldanha Bay area is busy with industry (the harbour exporting iron ore, the modern steelworks) and leisure resorts. In fact, its pristine beauty is fast disappearing under ghastly *faux* Mediterranean holiday housing and a visit is desirable to see it now (as soon as possible) before it vanishes forever.

But on a beautiful day when the sky is a deep blue, the breeze has dropped and the bay is calm, it is still possible to recreate in the mind the elegant sails, the sturdy ships, the flash of cannons and, in climax, the terrible explosion of the

Middelburg. Schapen Island hasn't moved and, on the deck of one of the hotels in the town of Saldanha Bay (with a beer in hand), overlooking Hoedje's Bay, one is within hailing distance of where the Indiamen rode peacefully at anchor before their rude awakening. The sea is undoubtedly the best place for the contemplation of the invisible past, especially on a warm day and over a cold beer.

If you drive on the road past the fish factory at Hoedje's Point towards the Port Authority control tower there is a turn-off onto a gravel road, with a rough metal sign indicating the path to the Namapotjie restaurant. Beneath the control tower is the site where the *Middelburg* went down. The road leads past the mussel farms to Marcus Island (which is sectioned off by a high wall to protect its bird colony from predators).

The view of the entrance to the Bay is best here. This is where the British fleet put the stopper in the bottle.

CHAPTER THREE

—•—

MUIZENBERG

I n February 1793 war broke out in Europe between France on the one hand and Britain and Holland on the other. France's first invasion of Holland was easily repulsed but a second in 1794 under General Dumouriez was entirely successful. It was helped by revolutionary fervour and republican sentiment in Holland, from which emerged a republican party (opposed to the royalists) called the Patriots. In January 1795 the Dutch government sued to Dumouriez for peace and the Prince of Orange fled to England.

With the outbreak of war the distant outpost of the Cape of Good Hope took on a very important strategic importance. For Britain its significance was that it overlooked the crucial sea route to India which her base in St Helena did not sufficiently cover. For the French, Mauritius and her Indian Ocean interests were seriously threatened by instability in the Cape. The fall of Holland in 1795 simply increased anxieties and intervention in one form or another became inevitable.

At the Cape itself, the head of government, Commissioner-General Abraham Josias Sluysken, was faced with an extremely difficult situation. The Dutch East India Company was already in turmoil, not helped by the war. The Colony, no longer backed by gold and silver coinage but reliant on paper money secured only by the Company's flimsy promises, was in an economic crisis.

In the interior the burghers were stirred by radical ideas such as the rights of man which came out of the American War of Independence and out of France before and after 1789. There was considerable discontent, too, about the Company's control of trade and the heavy taxes that prevailed. On the frontier the burghers complained of the constant harassment by Bushmen; they were incensed that the local landdrost in Graaff-Reinet did not take seriously enough nor reveal to government in Cape Town the extent of the occupation by the Xhosa in the Zuurveld and the immediate danger to their welfare.

Failure to move quickly enough to redress these grievances led to the overthrow of Company rule by the town of Graaff-Reinet (starting in February), followed by Swellendam (in June). Loyalties in Cape Town and Stellenbosch were also divided: some favoured the royalist Orange Party, others the republican Patriots.

The remoteness of the colony meant that in mid-1795 Sluysken was largely in the dark as to what was happening in Europe. In April 1795 the Dutch frigate *Medenblik* brought news that the French advances were considerable and serious, but it had left Holland before the collapse of the monarchy and the enthusiast alliance of Holland's republic with France. For all Sluysken knew France was still an enemy and Britain an ally: his duty was still to the company and to the States-General. While the threat of a foreign invasion was a distinct probability, Sluysken could not be sure whether he would welcome his compatriots or whether he would need to brush up his English or his French.

Prudently, he looked to his defences, both in mortar and in men.

When Van Riebeeck landed at the Cape he had immediately set about building a fort with earth walls initially seven foot high. This was certainly adequate to deter attacks by the local inhabitants even before the walls were heightened. But the fort was certainly no deterrent to any invading European force, especially when rain kept dissolving the mud walls, so, when war once more threatened in 1665, a new site was chosen a couple of hundred metres from the original fort. The new pentagonal stone castle, with walls at least 32 feet high took 13 years to complete. One hundred cannons prickled its ramparts like a porcupine.

But it was not long before doubts arose about the Castle's effectiveness and its safety. Its guns were there to overlook the anchorage in Table Bay but there were large areas beyond the range of their firepower; from Devil's Peak and what is now Signal Hill the Castle itself was vulnerable to enemy artillery.

So in 1715 the governor, Mauritz de Chavonnes, a military man, set out to remedy the first problem by constructing a battery at the foot of Signal Hill (where Albert Basin is now located on a site very close to the clock tower). It was built upon rock and its 16 cannon cast a formidable eye on anchorage beyond the range of the Castle. It was called the Mauritius Battery (the name was changed in 1744 to the Chavonnes Battery).

The coastline to the east of the Castle, however, remained completely open to land invasion so, in 1743, Baron von Imhoff set about fortifying the area. A string of batteries ('Elizabeth', 'Helena', 'Charlotte' and 'Tulbagh') were constructed – connected by a breastwork running along firm ground parallel to the sea – culminating in a strongpoint called Fort Knokke (located in the area now called Woodstock). The entire set-up became known as the Sea Lines.

Of course, the Sea Lines pointed outwards and were protection only from attack from the sea. They could easily be outflanked from the land.

In 1781, during the American War of Independence, Commodore Pierre André de Suffren had inflicted sufficient damage on Commodore Johnstone's British fleet in the Cape Verde Islands to delay it, and beat him in a race to the Cape. For the next three years, dancing to the dash and panache of its French visitors, Cape Town became little Paris. Before they left, the French commander, Conway, remedied the defects of the land defences with a line of fortifications running southward from Fort Knokke towards Devil's Peak. It ran up the slope ending at Zonnebloem Farm, and had three strongpoints called 'Holland', 'Centre' and 'Burghers'. These became known as the French Lines. The French also began work on a fortification – to be known as the Amsterdam Battery – between the Chavonnes Battery and the Castle – to protect the former from a land attack.

Save for the Castle the only one of these structures to survive down to the present day is the Centre Redoubt of the French Lines which can be found, together with one of the original cannons, in Trafalgar Park in Woodstock. It is a modest earth-walled fortification but is certainly worth a visit to see an example of these defences.

Hout Bay was also seen as a weak spot and two forts were built, one on either side of the bay, and a redoubt (called the Conway) was constructed at Constantia Nek. A battery was also set up at Kloof Nek.

Willing or nilling, Sluysken could not do much to improve his manpower defences. The main force consisted of an infantry regiment with 25 officers and 546 other ranks and an artillery corps of 27 officers and 403 other ranks. In addition there were 57 men of the depots of the Meuron and Wurtemburg regiments and a corps of 210 'Hottentot' (Khoikhoi) pandours. (The name comes from a Croation word which was transferred to the Khoikhoi regimental soldiers.) The infantry officers were fervent Orange men while the other ranks were more mercenary in their allegiances. The artillery men on the other hand were distinctly republican, favouring the French.

The infantry was headed by Lieutenant-Colonel de Lille and the overall commander of the military was the attractive and romantic figure of Colonel Robert Jacob Gordon (born in Holland of Scottish parentage).

After months of suspense a report finally reached the Castle on 11 June that a fleet of ships was beating into False Bay. At midnight Sluysken dispatched De Lille with 200 infantrymen and 100 artillerymen to reinforce the small garrison at Simon's Town. They arrived there at noon, not knowing quite what or whom to expect.

Admiral Nelson is still known and revered to this day (after all, his flagship

Victory lies, as a British national shrine, at berth in the Thames, dozens of English pubs were named after him and his name is even commemorated in a particular cricket score), but Admiral Elphinstone? How many would recognise his name to win top prize in 'Who Wants to be a Millionaire?' or 'The Weakest Link'?

Yet George Elphinstone, Viscount Keith, had a most distinguished career. He commanded a ship during the American War of Independence and was present at the capture of Charlestown. In 1793, as a captain in a naval squadron under Admiral Hood, he went head to head with the young artillery officer Napoleon at the Siege of Toulon.

Now, approaching 50 years old, he found himself in command of the British Task Force whose mission was to secure the South Atlantic. He dispatched an advanced squadron under Commodore Blankett on 13 March with some troops under Major General Craig. He himself followed with the main fleet on 3 April and a third force, carrying the main army, under General Sir Alured Clarke, set out soon afterwards. Blankett and Elphinstone were due to join up at St Helena but somehow missed each other, so they were first reunited off Simon's Town on 10 June.

The British probably chose to approach Simon's Bay partly because it was so lightly defended – there were only two batteries there, called Boetselaar and Zoutman, with just a few puny cannons – and partly because, unlike Table Bay, it provided relatively safe anchorage during winter.

To the few Dutch defenders the British fleet must have been a sight of terrible beauty. There were three 74 guns (the flagship *Monarch*, *Arrogant* and *Victorious*), three warships with 64 guns (*America*, *Ruby* and *Stately*), a frigate (*Sphinx*), and two sloops (*Echo* and *Rattlesnake*). Almost immediately, De Lille was ordered to pull back to Muizenberg where there was no port but a signal station and a couple of mortars.

A deputation of British officers handed Sluysken a letter from the Prince of Orange asking him to welcome the British as friends and one from Elphinstone and Craig which kept the Cape governor in the dark concerning the true state of affairs in the Netherlands. In a quandary Sluysken temporised but for a while agreed to provision the British fleet (theoretically they were still allies).

For nearly a month little happened though Sluysken's army was reinforced by over a thousand mounted burghers from Stellenbosch and even a detachment from the rebellious Swellendam. Finally, Elphinstone decided to act. On 9 July he seized three Dutch ships and on 14 July he landed 450 men of the second battalion of the 78th Regiment at Simon's Town. Even then not a great deal happened until 3 August when a burgher officer and a small detachment of pandours in the hills above the town fired on a picket, slightly wounding one of the British soldiers. In this desultory way the fight began.

Although not seriously threatened the British were not in a great position to launch an attack until the main army under Clarke arrived. They had no field pieces and could only scrape up some 1 600 men when marines and superfluous sailors were flung together with the regular infantry of the 78th Regiment. This was less than half the force available to Sluysken.

Furthermore there was only one road north to Cape Town which could easily be dominated by artillery. The key to the situation was Muizenberg. Here the mountain came down to within a few metres of the sea. The road – which follows closely the present road between Muizenberg and Simon's Town – swept round this curve between steep rock and sea. Beyond this the open countryside of the Cape Flats gave an obstacle-free approach to the Mother City. Craig would have to take the position, clear it and hold it. Not an easy task since the Dutch at the uncomfortable camp consisted of an 800-strong force of mixed regulars, artillerymen, burghers and pandours, soon to be joined by another 100 burghers. And 11 artillery pieces trained on the single road presented a formidable obstruction. They did not, however, believe they were at risk from the sea since the warships would not be able to get in close enough. Only two 24-pounders were placed facing the sea but they were not given a stable firing platform so their effectiveness was doubtful.

Craig and Elphinstone decided to make their move on 7 August. The British column began to advance some time after noon. Aside from having to attack on a very narrow front, the most pressing problem facing them was the lack of field guns. But the experienced Elphinstone took advantage of the peculiar nature of the terrain. He converted a small ship into a gunboat which he called the *Squib* and mounted an 18-pounder and a 9-pounder gun on it. Together with some armed launches the *Squib* was rowed along the shore a short distance ahead of the column. Further out, *America*, *Stately*, *Echo* and *Rattlesnake* bristled with deadly cannons, a moving platform of supportive artillery. The sea was calm and a gentle breeze whispered in from the north-west allowing the ships to move slowly and efficiently.

At Kalk Bay a scattering of shots from the fleet dislodged a picket which fled across the mountains. When the fleet reached Muizenberg it poured broadsides into the Dutch camp. This very quickly persuaded De Lille and the infantry to abandon the position and flee northwards through Sandvlei. One company under Captain Warnecke retired along the base of the Steenberg in more orderly fashion. The artillery followed except for the 24-pounders under Lieutenant Marnitz who replied to the fire of the warships. Two men were killed and a few wounded on board *America*, and *Stately* had a man wounded. But Marnitz soon realised his guns could not fire accurately; he spiked them and retreated as the British charge gathered steam along the road from Simon's Town. The Dutch camp was thus taken with ease.

However, as soon as they were beyond the range of the ships' cannons, the artillery and some burghers turned to make a stand and had to be dislodged by a charge of two companies of the 78th led by Major Moneypenny (whose female descendants may or may not in later years have joined the administration of British intelligence).

A second stand a bit later, backed by an artillery piece commanded by Captain Kemper, was more serious and forced the British to fall back on Muizenberg. Casualties were light on both sides.

De Lille was not seen again on the field that day but he did return to Sandvlei on the following day. He did not stay for long and fled, not stopping till he reached Wynberg where he encamped. The advancing British were, however, checked by a group of burghers and pandours hidden behind some sandhills, and forced to retreat a second time. Had De Lille been there to follow up this success things might have been more exciting for Craig and his troops.

Consequently, some burgher officers accused De Lille of treason and he was carted off to the Castle. His command was taken over by Captain van Baalen. (He was later acquitted of the charge.)

When the *Arniston* arrived from St Helena with a few hundred men and nine artillery pieces on 9 August the British position was strengthened and Elphinstone and Craig wrote again, demanding submission. The council of policy in Cape Town, by a large majority, rejected this and proposed to resist resolutely. So a stalemate developed over the next three weeks.

On 1 September a burgher force, backed by some pandours, attacked some vedettes on the Steenberg. In repulsing the attempt Captain Dentaffe of the St Helena regiment (the newest arrivals) was wounded. As was Moneypenny.

But as the British grew stronger, the Dutch weakened. Trouble from Bushmen inland and a threatened slave insurrection at Stellenbosch and Drakenstein siphoned off many of the burgher horsemen, who feared for their families. At the beginning of September, too, a large number of the pandours mutinied over a variety of grievances and, though they were pacified by Sluysken when they marched on the Castle, they were never again much of a fighting force in this campaign.

A plan was drawn up for a night attack on Muizenberg but it was too late to do anything about it. On 4 September an imposing convoy of transports sailed into Simon's Bay bringing Major-General Alured Clarke and the 84th, 95th and 98th regiments, together with detachments of artillerymen and engineers. For the Dutch the game was up.

On the morning of 14 September, two columns of the British, numbering between four and five thousand men, struck northwards from Muizenberg to Cape Town. They were harried effectively and courageously by a small force of burghers from Swellendam under Daniel du Plessis and lost one sailor killed and 17 soldiers wounded. But this was pinprick stuff: nothing could slow the juggernaut.

A very brief and badly handled stand by Van Baalen was made at Wynberg but was easily brushed aside. That evening the British army camped at Newlands. The Dutch forces still comprised about 1 700 men but the fight had gone out of most of them.

Capitulation was finalised on 16 September and a large column of British infantry arrived at the Castle that afternoon. The Dutch marched out with colours flying and laid down their arms in surrender.

Clarke, impressed by his resistance, invited Du Plessis to dinner. A couple of weeks later, Gordon, distraught by the abuse heaped upon him for his conduct and by the shame of the Colony's collapse, committed suicide. Sluysken sailed off to arrive eventually in Holland in April 1796. He was investigated by the National Assembly over his conduct in the loss of the Cape. Despite some embarrassing accusations against him, his name was cleared.

There is a suspicion that, because of their Orange sympathies, De Lille and Sluysken did not resist as strongly as they might have.

CHAPTER FOUR

SALDANHA BAY (2)

Many of the early battles in South Africa were really extensions of wars in Europe. Colonisation of the country was in many ways simply a by-product. The last direct threat to the first British occupation came from the Dutch, their erstwhile allies.

The armies of revolutionary France conquered the Netherlands, which became the Batavian Republic (in 1795) and a protectorate of France.

The new government of the Netherlands set about trying to protect its colonial possessions and establish its own legitimacy and principles in these territories. The trouble was that its fleet, strong on paper, was weak in practice. All it could muster were three ships-of-the-line – the *Dordrecht*, as flagship, the *Revolutie* (both with 64 guns) and the *Maarten Harpertzoon Tromp* (54); two frigates – the *Castor* (44) and the *Braave* (42); and three smaller ships – the *Bellona*, the *Sirene* and the *Havik*. An Indiaman, the *Vrouw Maria* (16), was to be used as a replenishment ship. They also took with them a small force of soldiers.

Rear Admiral Engelbertus Lucas was to take command. He was not an ideal choice. He had been to India but he had little command experience, less skill and negligible acumen. What he did have was what has bedeviled so many military campaigns of history – loyalty to the (Patriot) cause and in particular to the faction that espoused it. His appointment was clearly a political one. However, while the ships under his orders were up to scratch, the men were much less so. Many were Austrian and German draftees of dubious enthusiasm and a sizeable number of them, sullenly if mutely, still supported the Orange cause and their reliability was in doubt.

His main goal was to secure the Cape but if the British had already taken it (and the Netherlands government knew they had before Lucas left) then the task was likely to be beyond his relatively modest force. In this case he was to go on to Mauritius and try to persuade the French to help him in his endeavour.

His main hope lay in the negotiations between the government and the French. Could not a joint venture be undertaken? The French gave a cautious affirmative, promising a strong fleet of perhaps 8 large ships and 4 000 soldiers but with the unpleasant rider that, in view of the detestable state of their finances, the Dutch would have to cough up three million guilders to help finance the outfitting. Some of the money was paid upfront, some committed for the future.

Although the French force was far from prepared when Lucas left on 23 February 1796, he was given three rendezvous points at which he might meet them and he certainly expected to do so.

Had Lucas left a few months earlier the ultimate outcome of his expedition might have been different. The British had marched into Cape Town on 16 September 1795 and Admiral Elphinstone had taken most of his ships and General Clarke many of his troops on to India, leaving only the *America* and *Ruby* (both 64 guns) and the smaller *Princess* and *Star* (prizes taken from the Dutch) as guardships under the command of Commodore Blankett. The garrison was reduced to fewer than 3 000, with General Craig in overall charge of the Cape.

Craig set about strengthening his defences. In the interior most of the burghers of Stellenbosch and Swellendam had been pressurised into a grudging allegiance but further east Graaff-Reinet was in unhappy turmoil. He also considered Saldanha Bay as a potential staging point for a counter-invasion and sent Blankett to examine it. The Commodore reported that the Bay presented a huge problem for any enemy exploitation of it because of its serious lack of fresh water. Craig therefore discounted it as a threat but posted observers there to inform him of any incursion by means of a relay of horses.

The Dutch fleet left Texel in high secrecy on 23 February 1796 and set a course around the north of Scotland and west of Ireland. But as it approached the Canary Islands a French-looking ship flying French colours was espied very close. When Lucas replaced the English flag he was flying with the Dutch flag, the unknown ship took to its heels. It was, in fact, the *Moselle* (18 guns), a French-built ship captured by the British. 'Let him go to the Devil,' said one of the Dutch captains.

Which was exactly what its captain Charles Brisbane did, if you think of the Cape in April as hell on earth. When the *Moselle* arrived at the Cape on 20 July, Craig was given plenty of advanced warning about what was approaching. (But not very fast.)

Lucas's fleet, delayed by storms, oppressed by sickness and riven by dissent, dawdled in the Canaries. Not so the British. The Dutch movements round the British Isles had been observed and the alarmed British wasted no time in sending reinforcements to the Cape. On 6 March, the *Sceptre* (64 guns), the *Crescent* (36) and six transports left Portsmouth with 1 500 soldiers; on 11 April *Jupiter* (50) sailed with a convoy conveying 2 400 troops; and the next day *Tremendous* and *Trident* (both bristling with 74 guns) followed with 2 000 more.

Lucas did spot the *Jupiter* and its charges on the horizon but let it go. On 14 May the lone *Tremendous* was also sighted but again was allowed to go unmolested. An aggressive admiral like Elphinstone would never have allowed such opportunities to slip through his fingers: Lucas was no Elphinstone.

In fact, that worthy commander had returned from India to Table Bay on 23 May with the *Monarch* (74) and the *Sphinx* (24). He was soon joined by the *Sceptre* and *Crescent* and, weeks after, by *Tremendous* and *Jupiter*. Elphinstone had been led by the *Sceptre* to expect that a combined Dutch-French fleet was on its way, so he was concerned. Lucas meanwhile headed first for Porto Praia (he reached it on 26 May but found there neither friend nor foe), then on to Rio where he anchored at the end of June. Again, there were no French there. So he started to cross the Atlantic, anticipating his rendezvous with the French fleet at last.

Lucas had been deceived. A French squadron under Rear Admiral Pierre Sersey had left Rochefort but it consisted only of *La Forte* (50 guns), *La Vertu* and *La Seine* (40) and *La Regenérée* (36), nothing like the eight ships-of-the-line that had been promised. In fact, the Dutch money had been diverted to a fleet engaged in preparation for an invasion of England! Even the squadron that was sent went with vague and ambiguous orders.

Intelligence in military matters is not just a state of mind. In the form of material information it is the irreplaceable staple that the commander feeds on. Sometimes it is stumbled upon, chance or luck prepared to show its hand.

The French squadron had made much better time than the tardy Dutch. On 26 May it captured the *Sphinx* east of False Cape. Previous to that, on 15 May, in the middle of the Atlantic it had surprised the whaler *Lord Hawkesbury*. The French crew put a prize crew on board but left two of the English sailors and a cabin boy with it. After a while the English sailors, on their best behaviour, were allowed to participate in the normal sailing of the ship. So, on 26 May, Edward Morrow was at the helm when the *Lord Hawkesbury* approached Cape Agulhas. Surreptitiously he headed the whaler for the shore. It ran aground off Zoetendal's Vlei where its French crew were taken prisoner and Morrow and his companions were able to inform Elphinstone that the frigates were all there was of the French fleet and that even they were on their way to Mauritius and not there primarily to link up with the Dutch.

Therefore, Elphinstone knew that he had significant superiority. He did, however, make a mistake, though it did not cost him. He wrongly surmised that the Dutch, knowing that it was now beyond their capacity to take the Cape, would bypass it and head for the east. So he went in search of them, wasting a fruitless week beating around the tempestuous ocean and leaving Craig undefended. When Craig got news that Lucas had anchored in Saldanha Bay on 6 August he tried unsuccessfully to send word to the English admiral.

Elphinstone returned on 12 August but bad weather prevented him from leaving Simon's Bay until three days later. His force was formidable: eight ships-of-the-line (*Monarch, Tremendous, America, Stately, Ruby, Sceptre, Trident* and *Jupiter*) and six other warships (*Crescent, Sphinx, Moselle, Rattlesnake, Echo* and *Hope*).

Craig had the luxury, by this stage, of being able to leave 4 000 troops under Major-General Doyle to guard the peninsula while he marched on Saldanha Bay with 2 500 soldiers and 11 field guns (this force included elements of the 25th Dragoons, the Grenadiers and the 78th and 80th Regulars). Lucas was to be assailed by land and sea.

When Lucas first arrived in Saldanha Bay he stationed *Havik* at Hoedje's Bay and *Bellona* at Langebaan to protect the precious water sources there. He put up tents on Schapen Island for the hundreds of sick, and the bulk of the fleet he took deep into the Bay. He put ashore three lieutenants to announce his arrival to the burghers and to procure intelligence. One of these, Valkenburg by name, was married to the daughter of a local farmer whose farm was about 15 miles away. There Valkenburg learnt that the Dutch could expect no assistance, let alone uprising, from the burghers, that the Cape was now too heavily protected for recapture and that the advance guard of an approaching British army was already in sight heading northwards.

Lucas was in trouble even before the British army and fleet arrived. In the Canaries there had been unrest on board the *Dordrecht, Revolutie* and *Castor* and the cry of '*Oranje boven, weg met de patriotten!*' had been heard and only with some difficulty suppressed. Now desertion was growing from a trickle to a haemorrhage. So Lucas made the decision to leave for Mauritius on the 16th.

He was too late. That morning the advance units of Craig's army reached the heights at Langebaan. The *Bellona* began a steady fire on them, though without significant effect. The British could not reply until a howitzer was brought up, forcing the *Bellona* to back off. Lucas dispatched two boats to evacuate the sick from Schapen Island but one of the boats forthwith defected to the British! The sick were consequently abandoned to their fate.

At one o'clock in the afternoon *Havik* signalled from Hoedje's Bay that there were a considerable number of sail in sight. By four o'clock the *Crescent* appeared at the entrance to the Bay. As Jose Burman and Stephen Levin say in their lively account of the engagement, 'The cork was being inserted in the bottle.' Elphinstone entered the Bay in double line and towards sunset anchored just beyond reach of the Dutch guns. The Dutch were in the eastern part of the lagoon, edging towards Schapen Island.

Elphinstone sent the aptly but ominously named Lieutenant Coffin on a boat under a flag of truce to Lucas's flagship. Coffin delivered a letter which was unequivocal: 'It is unnecessary for me to detail the force I have the honour to command, because it is in your view and speaks for itself, but it is for you to consider the efficacy of a resistance with the force under your command. Humanity is an incumbent duty of all men, therefore, to spare an effusion of blood, I request a surrender of the ships under your command, otherwise it will be my duty to embrace the earliest moment of making a serious attack on them, the issue of which is not difficult for you to guess.'

Lucas was caught between a rock and an Elphinstone. He asked the British admiral for time to consult his officers and give a reply in the morning. Elphinstone acceded to this only after receiving a written promise from his Dutch counterpart that he would not scuttle his ships. Lucas met with his officers that night.

At nine o'clock next morning Captain Claris boarded *Monarch* with Lucas's terms of surrender – that the Dutch be allowed to proceed to Holland on the *Braave* and the *Sirene* with their possessions, whereupon these ships would go on to England. In the light of the fact that such pledges had not always been honoured by the authorities in ports controlled by the French, Elphinstone refused and agreed only to the retention of private property by everyone and to the transport home of officers in British ships provided they pledged not to serve against the British as long as hostilities lasted. The crews could choose between volunteering for service in the East India Company or being pressed into the Royal Navy.

At five in the afternoon Valkenburg brought the written acceptance of Lucas aboard *Monarch*. At the same time he urgently requested that the British take possession of the *Dordrecht*, *Revolutie* and *Castor* because the crews were threatening the officers, trampling upon the republican flag and were now shouting '*Oranje boven, de dood aan de patriotten!*'

Most of the sailors were happy to change sides and the Dutch ships, in excellent condition, became a valuable addition to the British navy. The *Castor* was symbolically renamed the *Saldanha*.

The Dutch officers were returned to their native country and a court was appointed to try them. Lucas was kept in solitary confinement. To the relief of everyone, including perhaps himself, he died of 'mortification and anxiety' but otherwise natural causes, on 21 June 1797, before any verdict could be handed down. The rest of the officers were acquitted.

Other than the brief action involving *Bellona* no shots were fired in anger during this second battle of Saldanha Bay. The victory was bloodless yet it was comprehensive. An entire and powerful fleet capitulated.

CHAPTER FIVE

—◆·◆—

BLAAUWBERG

On a summer's day in January, hot, torpid, bright, one of the best places to be for breakfast is the Big Blue Café in Melkbosstrand on the West Coast. You get a generous helping of bacon, eggs, sausage, tomatoes, banana, apple, fried onions and mushrooms, all at a reasonable price.

Over the steam of your tea, over the spray of the breakers, over the haze of the waves, over the distance of 200 years you can let yourself imagine a huge fleet with sails and pennants and figureheads imperiously moving into sight and taking up position offshore. One large 64 gun man-o-war (*Diadem*), one 32 gun ship (*Leda*) and two brigs (*Encounter* and *Protector*) dropped anchor as close to land as possible in order to cover the landing-place with their heavy cannons. Over to the left is the most distinctive feature of the bay – a jagged reef of sharp rocks that provides the shoreline a measure of protection from heavy swells. A small ship was run aground there to form a breakwater for the invasion about to come.

You can imagine, behind you, the few skirmishers the Dutch had sent to watch. You can imagine what went on in *their* minds as they realised that this was to be one of the largest seaborne invasions Britain had ever undertaken. It was 6 January 1806.

The Cape had once more become strategically important when Britain declared war on France in May 1803 after it had been returned to the Dutch earlier that year.

This time the Dutch were unavoidably attached to the French cause since the French, following their 1794 to 1795 invasion, had set up the Batavian Republic in 1798. In 1806 what was now the Batavian Commonwealth was represented at the Cape by Lieutenant-General Janssens.

Janssens' situation was unenviable. His best regiment had been ordered to Batavia itself so that his military strength, never sufficient, was now dangerously depleted.

The overall population of the Cape at the time consisted of about 25 000 whites, some 20 000 'Hottentots' in service and 29 000 slaves. Under arms Janssens could call on a coloured (so-called Hottentot) regiment of between 500 and 600 strong under Lieutenant-Colonel Frans le Sueur; the 5th Waldeck Battalion consisting of about 400 German and Hungarian mercenaries; the 9th Battalion of jaegers (infantrymen, often hunters or game wardens renowned for their markmanship) some 200 strong, recruited from all over Europe; and an odd assortment of Batavian marines, dragoons and mounted burghers. From the French ships *Atalante* and *Napoleon* came 240 marines under Colonel Guadin Beauchêne who was the marines' commander on the latter ship. He had some 16 field guns manned by 54 Javanese artillerymen called Mardykers (a name derived from Campon-Mared, an area in the East Indies). One hundred slaves also helped in the moving of the artillery.

Given the strategic importance of the Cape as the only sea route to the East, Janssens had every reason to expect some form of intervention from the British.

When war had been declared Napoleon had prepared to invade his island enemy with some considerable intent. He brought his *Grande Armée* to Boulogne and assembled some 2 000 ships between Brest and Antwerp for the invasion of the island. He needed to control the Channel for just a brief while but the British fleet kept him at bay. Even an alliance with Spain and the entry of the Spanish fleet on the French side failed to correct the imbalance. Nevertheless the British felt that the Cape must be secured.

Quietly, with considerable subterfuge, they assembled a very large army and a fleet of over 60 ships. In July 1805 at Falmouth, the 59th Infantry Regiment, the 20th Light Dragoons, 320 artillerymen and assorted recruits embarked in transports under the protection of the brigs *Espoir*, *Encounter* and *Protector*, supposedly destined for the East Indies. Not long thereafter the 24th, 38th, 71st, 83rd and 93rd infantry regiments boarded more transports at Cork. They were escorted by *Diadem*, *Raisonable* and *Belliqueux* (all 64 guns), *Diomede* (50 guns), and the frigates *Narcissus* and *Leda* (32 guns).

The fleet was under the overall command of Commodore Sir Home Popham, the army (over 6 500 troops) under Major-General David Baird. Word was spread that they were headed for the Mediterranean. Since the threat of invasion from Boulogne was still real this force seemed a sideshow of relatively little importance.

In fact, while they were at sea events of crucial import occurred to affect the course of the war in Europe. On 21 October 1805 Admiral Nelson caught Villeneuve's French fleet off Cape Trafalgar and dispelled once and for all – at the cost of his life – the fear of invasion. But even before this Napoleon (like Hitler in the following century) had turned his attention to the east, crushing the Austrians at Ulm in October and capturing Vienna in November. Then in December he won a crushing victory over the Russians and Austrians at Austerlitz.

None of this was known to the British fleet headed for the Cape. They reprovisioned at Madeira and St Salvador (where they bought horses for the dragoons) and ran into trouble off Brazil where two transports were wrecked with the loss of Brigadier-General York. The frigate *Narcissus* and the brig *Espoir* were sent ahead on separate missions to gather information. The latter managed to glean details from a neutral merchantman about the strength and disposition of the Cape garrison.

General Baird had had an exciting and distinguished career, much of it in India. In 1780 he was wounded at Perambaukum and held a prisoner in irons until his release in 1784. He was promoted to lieutenant-colonel of the 71st Regiment in 1790 and took part in the second war against Tippoo Sahib which ended in 1792. In 1799 in the third war against the Sultan of Mysore, at the head of the Highland Brigade, he led the attack which stormed the defences of Seringapatam where Sahib himself was killed in the breach. In 1801 Baird and his army landed on the Egyptian shore of the Red Sea and marched across the desert to the Nile. They then converged on Alexandria, but the port had fallen before he arrived.

There was something else in his favour – he had spent a year at the Cape during the first British occupation and had first-hand knowledge of its defences and its terrain.

Ironically, these defences had been strengthened by the British as soon as they had taken over Cape Town in 1795. General Craig had learnt from his own success. The French Lines had not been carried through to the slopes of Devil's Peak so he constructed three blockhouses (called the 'York', 'King's' and 'Prince of Wales') to remedy the deficiency. The King's Blockhouse was located high up on Mowbray with an excellent view of both Table and False bays. It is easily visible from many parts of the city's suburbs.

Between Fort Knokke and Salt River the rather feeble Nieuwe Battery was replaced with a stout square blockhouse from which an old cannon and five 24-pounders could cover the sea and the land. It was known as Craig's Tower. Since the British had built it they had a healthy respect for it. Baird had no doubt seen it and admired it; now it stood in his way and he might have to knock it out!

Baird had the option of landing in False Bay or Table Bay. In fact it was no real option at all. Since the strengthening of the defences Simon's Town and Muizenberg would be an even more difficult approach. Table Bay it had to be.

Janssens was on tenterhooks. Intelligence at sea was, of course, very slow in those days. But on Christmas Day his worst fears were confirmed.

A French privateer, the *Napoleon*, ran itself aground on the coast south of Hout Bay. The day before she had been spotted by *Narcissus* which had immediately given chase. Being run ashore was the only alternative to capture or sinking. Where there was one British warship there were likely to be more. This was almost immediately confirmed by the arrival of another ship which reported that it had passed a large convoy on its way south. On the 28th came a further account of a large British fleet sailing from Madeira on 3 October. On 4 January 1806 the wait was over. Sentries on Signal Hill saw the horizon begin to fill with sails and by afternoon the enemy fleet, awesome in its magnitude, had moved into Table Bay and taken up position between Robben Island and Blaauwberg.

Signal guns sited on a line of hills relayed the news of the British arrival so that within eight hours of the first sighting the burghers of Swellendam had been alerted. But the journey was long, the heat during the day intense and the farmers reluctant to leave their ripening crops. They would not play a significant part in the coming battle.

After an abortive attempt to put part of his force ashore at Camps Bay, General Baird's plan was to land his troops next morning on the beach of what was then called Losperd's Bay (now Melkbosstrand). It was sufficiently far from Cape Town for him to expect a relatively unopposed, at worst a lightly opposed, landing and a 26-kilometre march to the town. But the plan was thrown into confusion when a south-easterly gale blew up overnight and a heavy surf next morning made a landing too hazardous.

Instead he decided to send part of his army north to the more sheltered haven of Saldanha Bay, intending to follow with his main force. So with *Espoir* in the lead, transports carrying the 38th regiment, the dragoons and part of the artillery headed north during the night of the 5th under the command of Brigadier-General Beresford and under the protection of *Diomede*. Baird was not happy with this scenario: Saldanha Bay was further away and the march south would be even harder and hotter with little water on the way.

On the morning of the 6th, however, the wind, though still stiff, had subsided sufficiently to make a landing at Losperd's Bay possible, if risky. The fleet took up its position accordingly.

At around noon the Highland Brigade, commanded by Brigadier-General Ferguson, and consisting of the 71st, 72nd and 93rd regiments, making up the first wave, assembled in their boats and headed for the shore. Ferguson had already gone ahead to scout the beach and pronounced it free of defences.

The morning, a Monday, was hot and assured of getting hotter. The Highlanders were in full uniform with heavy loads of equipment. Colonel John Graham of the 93rd (the Argyll and Sutherland Highlanders) was one of the invading party and described some of the confusion which accompanied it: 'The boats unavoidably got into such a crowd that many of the turning boats could make no use of their oars. We were then not a half a common shot from the beach or sand hills. One of the boats of the *Charlotte* had been unable to push to

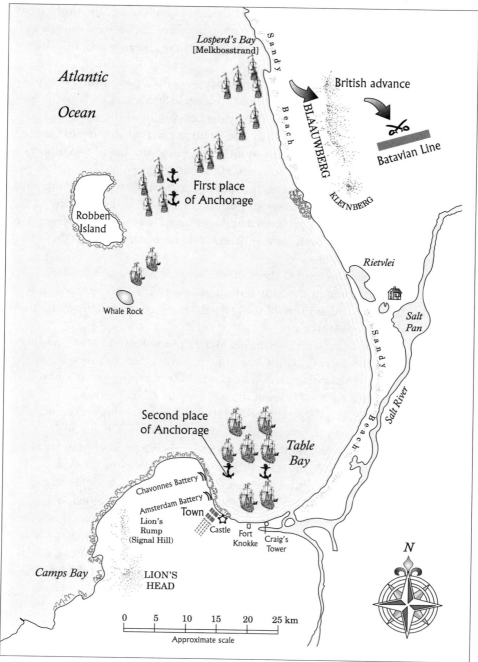

The Battle of Blaauwberg (based on a sketch attached to the contemporary report of Sir Home Popham)

windward of the rock, she touched it, instantly turned bottom up, and down went 36 of our brave fellows, cheering as they sank. Only three bodies were thrown up on the shore, or ever seen again.' What exactly they were cheering about Graham did not explain.

As the other boats approached the shore they were fired upon by the muskets of a small patrol of coloured soldiers under Colonel le Sueur sent by Janssens to harass the landing. They were kept at bay by the heavy guns of the warships and the landing was effected with the loss of one Highlander (probably James Conrie who died of his wounds the following month) and the wounding of two officers and three men. Le Sueur also had two men killed and three wounded.

But the wind had strengthened again, making further boat trips too dangerous. This left the Highlanders in a precarious position that night. If the weather worsened and they were not provisioned and reinforced they would be vulnerable to attack. There must have been some pinched and taut nerves amongst them on this foreign shore that night.

Next day, however, was kind to them, stores were got ashore and the 83rd regiment, the 1st battalion of the 24th and a battalion of the 59th joined them. Lieutenant-Colonel Baird, brother of the general, was in command. By nightfall the whole army was ready to move.

Janssens was presented with a dilemma. Many expected him to play a waiting game, marshalling his defence on the line of forts constructed for just such an attack. Instead he pushed northwards with about 2 000 men (one source has 3 000) to confront his enemy. He spent the night of the 7th at Rietvlei.

Few reasons have been suggested for this strategy. Janssens might have wanted to catch the British before they consolidated and somehow pick them off in scattered groups. In fact, had he arrived a day earlier he might have been able to attack the Highlanders while the rest of the army looked on from the ships.

For the British the most distinctive feature in their immediate sights was Blaauwberg mountain. It is 231 metres high. To the south it was joined to a smaller hill called Kleinberg; to the east a not very formidable ridge ran away towards a great plain. As they moved south the British would have to circumvent Blaauwberg either to the left or the right. The coastal route to the right was probably considerably less suitable, as the sand dunes and short but thick fynbos was not suitable for columns to march through, nor the soft sand easy to drag cannons over. So the army headed for the ridge to the left, sailors having to haul the artillery (consisting of two howitzers and six field guns) over the heavy terrain and up the ridge. Several of the sailors were to die of heat exhaustion during the course of the sweltering day. They started their advance at three o'clock in the morning of the 8th.

Janssens had arrived at Rietvlei at six o'clock in the evening of the 6th. There he made his camp. In late afternoon of the 7th several warships began a bombardment of the camp from the sea which they kept up for several hours. But Janssens had already left, moving some kilometres to the north, where he took up a position on the plain east of Kleinberg. On Blaauwberg itself he posted a small piquet to observe the movements of the British and on Kleinberg he established a small force of mounted burghers and a cannon as a cover for this left flank. At three o'clock on the 8th he roused his troops. The two armies began to converge for a battle which was now inevitable.

Benign from a hearty breakfast in Melkbosstrand the satisfied battlefield tourist can take two routes south – both pass to the west of Blaauwberg mountain. The more inland route gives a fine view of the approach to the mountain and the ridge from the British perspective but the coastal route is prettier and gives a better feel for the difficulties of the terrain on the west side of the mountain.

Either road gives easy access to the twin boroughs of Bloubergstrand and Table View and on the seashore you will find an excellent information centre run by the friendly, energetic and imaginative Pat Gee who can fire your enthusiasm for the many and varied local attractions. Through her I came to know Pat Matejek who is equally passionate about each and every plant and animal in the area.

The next place to visit in one's search for the battle of Blaauwberg is Rietvlei, where Janssens camped on the night of the 6th. This you can find by taking the Blaauwberg road – the main road through Table View town leading off from Marine Drive. Within a few hundred metres turn into Pentz Drive and follow this till you come to the world-famous South African Foundation for the Conservation of Coastal Birds (SANCCOB) – renowned most recently for its vast effort to rescue and restore to health thousands of oiled penguins.

In the car park outside SANCCOB you will see an interesting plaque on a wall opposite. On this spot, the site of a late Stone Age settlement, a dairy was operated on behalf of the governors of the Dutch East India Company by, amongst others, Wolraad Woltemaade, the man who, repeatedly plunging into the sea on his horse, saved so many lives when the *Jonge Thomas* went aground in a storm in 1773 and eventually lost his own life in the process. This is the general area where Janssens' camp was located and a few metres walk down towards the vlei you will find a beautiful spring (somewhat overgrown with reeds but alive with birds) which was probably the reason why the Batavian army paused there.

On the 7th Janssens moved his forces about 14 kilometres north to Blaauwbergvlei to the east of Kleinberg with his left flank on its slopes. Why did he take up this position? There were probably two reasons. Firstly, he now straddled the wagon road leading to Cape Town down which he might have expected the British to march. Secondly, there was another natural spring there which he would have wanted to deny to his enemy. It is possible to visit this spot but it is on private property and can be done only with prior permission.

The best place to view the whole battlefield is on top of Blaauwberg itself but it is not accessible to ordinary cars and again permission is needed. The two Pats went out of their way to arrange this for me and Pete Reinders volunteered to take me. It is typical of the friendly nature of the Table View/Blaauwbergstrand community that Pete's initial kind offer was soon joined by others so that late afternoon saw a small fleet of 4 x 4s battle its way up Kleinberg and onto the top of Blaauwberg itself. Gin and wine and champagne flowed, snacks and exotic breads and chicken were produced. This festive gathering in no way detracted from or undermined the fantastic vantage point this spot provided to view the whole battlefield and the vulture's eye-view it gives.

To the north is Losperd's Bay (Melkbosstrand) and in the distance is Koeberg nuclear plant. To the south is Table Mountain and the objective of the British advance, Cape Town. To the north-east lies Koeberg Hill and at a distance to the south-east is the line of the Tygerberg Hills. Down below, on the eastern side, lies a large and beautiful plain stretching all the way to the city. It was here on the eastern side of Blaauwberg, on the plain, that the main battle actually occurred.

Janssens drew up his force on a front of some 1 600 metres, with the mounted burghers and some artillery on Kleinberg and its slopes on the west. Next to them was the coloured infantry regiment, barefooted and dressed in white pantaloons and short blue tunics and high glazed black hats, ringed with white leather straps, pipeclayed and embellished with white feathers. Then came the Waldeck jaeger battalion. Next to them in the centre was the 5th Waldeck infantry regiment, then the French marines and on the right the 22nd Batavian infantry and the 9th Batavian jaeger infantry and more artillery manned by the Mardykers covering the flank.

Baird believed Janssens had about 5 000 men available to him at the Cape so he considered the odds to be even. But he must also have known that the real difference lay between the mixed and debatable training of his opponents and the hard-bitten discipline of his own troops.

Baird divided his force into two brigades which marched side-by-side over the ridge and descended into the plain. On his right up against the mountain the first brigade commanded by his brother consisted of the 83rd, 24th and 59th regiments; on the left was the Highland brigade under Brigadier Ferguson comprising the 71st (City of Glasgow regiment), 72nd (the Seaforth Highlanders) and 93rd. Baird's first action was to dislodge the observer scouts from Blaauwberg itself.

Janssens rode up and down the line encouraging his troops, cheered by all except the Waldeck battalion, cynical mercenaries that they were.

Action started when grenadiers of the 24th were sent to dislodge the Dutch right flank on Kleinberg. This they did with the loss of Captain Foster and three other men while sixteen were wounded and three went missing. These were the first losses the 24th took on South African soil: they were to lose many more in the future.

Then the Highland Brigade extended its line to face the whole of the Batavian front, pushing the first brigade into reserve. They were an impressive sight – shoes, not boots; red and white hose; small bonnets with feathers; large sporrans; red coats with buff or yellow facings; kilts of MacLeod, McKenzie or Sutherland tartan; black cartouches in the middle of their backs with 50 musket balls for the unrifled Brown Bess flintlocks. The artillery on both sides, which were now some 1 800 paces apart, was beginning to get uncomfortably close.

Captain Carmichael noted that at this time the Highlanders 'continued to advance over a tract of ground where we were buried up to the middle in heath and prickly shrub'. Their first volley of musketry was, however, fired at too great a range. They then fixed bayonets (and their bagpipes) for a charge!

Janssens may have placed the Waldeck battalion in the centre because he regarded them as the toughest and most experienced. He intended to hold fire until the Highlanders had discharged their muskets. Ironically, when a few cannon balls kicked up dust amongst them, the Waldeckers were the first to give ground in spite of Janssens' personal pleas and exhortations to consider the honour of their native land, Germany. On their left the coloured soldiers stood firm, as did the French sailors on their right, and the Malay gunners bravely kept up their fire. But when the Waldeckers saw the Highlanders charging towards them they turned in confusion and fled. Then the 22nd Batavians also began to buckle.

The marines tried valiantly to plug the gap but Janssens saw it was futile and ordered a general retreat. Adjutant-General Rancke and Colonel Henry were despatched to try to rally the fleeing troops at Rietvlei. A company of mounted artillery under Lieutenant Pelegrini carried on firing to the end and were the last to leave the battlefield. The governor was so impressed that he promoted the lieutenant to captain there and then.

Possibly it was the 71st and 72nd on the British side that led the charge and took the brunt of the fire because, according to a list of casualties, the former suffered 16 dead and the latter 6, but the 93rd, which had had 36 drowned two days before, lost only 3 on the day of the battle.

Janssens' shattered army collected together at Rietvlei. His losses were 110 of the (French) marines, 188 of the regular soldiers, 4 burghers, 17 coloureds, 10 Malays and 8 slaves. Not all of these were killed – some may have fled, never to be seen again. Janssens had well before the battle made a contingency plan to retire to the Hottentots-Holland mountains where he had arranged a reserve of provisions. He sent the disgraced Waldeck regiment back to Cape Town. One company which had conducted itself with honour in another part of the line he offered the choice of accompanying him, which they accepted. The gallant French he was reluctant to part with but Colonel Beauchêne suggested that they had little practical future at the Cape.

Janssens also sent instructions to Major Horn at Simon's Town to scuttle the warship *Bato*, spike the guns in the town and join him in the mountains. He himself followed the remnants of his army which rapidly made its way to the Hottentots-Holland via Rooseboom. That evening the British occupied Rietvlei, spending the night under the stars.

Baird's total loss had been 1 officer and 14 other ranks killed, 9 officers and 180 other ranks wounded and 8 ORs missing: not heavy enough to stop his advance on Cape Town the next morning. The fleet, too, moved from around Robben Island to anchorage just opposite the Castle, and put ashore a battery train which joined Baird at Salt River.

Lieutenant-Colonel van Prophalow, who had been left with a force of burghers and soldiers to man the forts and protect Cape Town, had no stomach for a fight, however, and sent a flag of truce to plead for cessation of hostilities for 48 hours while he prepared to capitulate. Baird gave him only 36 hours and insisted Craig's Tower (for which the British had such a healthy respect), Fort Knokke and the defensive lines be given up within six hours. Beresford and his force from Saldanha, after a gruelling march of some 150 kilometres in wasting heat, arrived at Salt River sometime after the 59th regiment moved into Fort Knokke. A march in vain!

On 10 January at 4 o'clock in the afternoon the capitulation was signed under a tree in Woodstock.

Baird moved quickly in the next few days to occupy Stellenbosch and sent the 83rd regiment by sea to take Mossel Bay. Janssens had considered denying the British access to the eastern districts and to try to sit things out in the hope that the French might send a relief expedition. (Indeed, on the 13th a French frigate, *La Volontaire*, sailed into Table Bay and was deceived into being captured by the British showing false French and Batavian colours.) But he soon realised the futility of this and on 18 January he signed the whole settlement over to the British. By the end he only had 659 soldiers and a few officers with him and these were allowed to sail for Holland. On the other hand many of the coloureds were induced to join the British and this became the nucleus of the Cape Regiment, and some of the Waldeckers also joined the British.

Janssens himself, with some of his government officials, sailed for Holland on the transport *Bellona* on 6 March.

Commodore Popham sailed off to attack Spanish interests in South America. This turned into a complete disaster and he was court-martialled.

The view from the top of Blaauwberg is stunning. Late in the afternoon we who were lucky enough to be there could see clearly the long plain stretching towards the city. Already in the distance the front line of houses is beginning to encroach on the plain with that relentless greed that characterises urban life. Human habitation oozes up the plain like an oil slick in a celestial sea or a pestilential weed in an enchanted garden. This heritage site with its buck and its birds, its mongooses and tortoises will soon be lost and the view vandalised by those too short-sighted to appreciate it. Pat Matajek is particularly concerned for the hundreds of tortoises which die every time that line inches forward. The site of the battlefield is in real danger from this and from the thoughtlessness of developers – the site of one of the most important battles in South African history. See it while you still can for soon it, too, will be just a memory.

In an ideal world it should be preserved as a tourist delight. There can be few battlefields in the world where the site is so clear and the view so stunning. The top of Blaauwberg provides a real-life panorama, a gem of a battle scene: the ridge over which the British made their way; Kleinberg; the plain where the Dutch drew up their line and from which the 24th dislodged them; and the rough terrain where the Highlanders made their decisive charge. Somewhere down there, in a site unknown, are buried in their uniforms where they fell some of the British dead. A European battle, fought by 13 different nationalities!

For our party, mellow from the food and drink, the light began to fade. To the west thin, low cloud, swirling mist, like a duvet, obscured Melkbosstrand and Koeberg nuclear power station in the distance and rendered Robben Island a barely distinguishable outline. But the summit of Blaauwberg stood in clear sky above the clouds like one of those American movie sets portraying heaven. To the south Table Mountain was sharp and clear and below it the city lights were bright. The view from Blaauwberg provides the best of one of the most beautiful sights in the world.

On our way down in the darkness we got lost in the maze of tracks and half-expected to come across the wandering figures of those three soldiers who went forever missing from the 24th.

CHAPTER SIX

——•—

SLAGTERSNEK

There is something odd about the Slagtersnek national monument (located on the main road – the N10 – between Port Elizabeth and Cradock, 9 kilometres south of Cookhouse).

Not only is it not at the Nek itself, but, as you walk the few metres to it from your car, it presents you with a blank gaze. You have to walk round to the other (eastern) side to find the inscription recording the hanging of five men at that spot on 9 March 1816. This is because the main road, which used to run in front of it, now passes behind it. Indeed, taking one's cue from this one might say that there is something uncomfortable about the whole episode.

The roots of the 1815 rebellion lie in a previous uprising in 1799. Then, Commandant Adriaan van Jaarsveld, who had been the scourge of the Xhosa in the First Frontier War, was arrested for forgery but freed by fellow conspirators such as Marthinus Prinslo and Coenraad de Buys (then living at Ngqika's Great Place across the Fish River, and 'sleeping with' the Chief's mother). The rebels were keen to distance themselves from the recent British occupation of the Cape. The acting governor at Cape Town, General Dundas, acted swiftly and sent a detachment of dragoons overland and two companies of the 91st Regiment (the

Argyle and Sutherland Highlanders) by sea – all under the command of General Thomas Vandeleur – to stifle the insurrection. Which they did very easily and very quickly.

Many of those who had taken part in this activity were, by 1815, living in the area called Bruintjieshoogte, at the eastern end of the plains of Camdeboo and just west of the Little Fish River. (In 1815 Lord Charles Somerset started Somerset Farm at the foot of the Bosberg to produce fodder for the cavalry horses and this later became the town of Somerset East.) Some located themselves to the north round Swagershoek while others pushed across the Great Fish River along the Tarka River or into the valley of the Baviaansrivier.

However, after Colonel Graham had swept the Xhosa out of the Zuurveld (that area between the Gamtoos and Fish rivers, also called Albany) in 1812, the Fish became the formal boundary between the Colony and the Xhosa. As a result no more farmers were permitted to settle to the east of it and those already there were not allowed to extend their farms. The farmers' sons, who only came of age at 25, could not therefore acquire their own farms and had to lead a potentially disgruntled existence under the patronage of their fathers.

The farmers chafed, too, under the (to them) unfair burden of the quit rent system (an annual tax on the estimated value of rented land). And they muttered darkly, after the Caledon proclamation was passed in 1808 giving the Khoikhoi some minimal legal rights, that the 'Hottentots' were being unduly favoured at their expense.

Johannes Jurgen Bezuidenhout and Cornelis Frederik Bezuidenhout were born in Tulbagh: Hans in 1758, Freek in 1760. They joined their younger brother Gerrit in the Baviaansrivier valley in 1809. Hans was married to Martha Faber and their oldest son Gerrit was christened just seven days before they were married. After being widowed Freek had taken up with a coloured woman and in 1815 he was living with another, Maria Eckard, by whom he had a small daughter.

The brothers were fiery men, quick to take offence and aggressive. Their eldest brother had killed the sister of Chief Malgas and Hans himself had shot dead the son of a Xhosa chief. They led a toilsome life, running a few cattle, cutting some timber and poaching wildlife. Hard men in a hard land.

In June 1813, an archetypal master-servant spat broke out between Freek and a Khoikhoi labourer named Booy. The latter, on completing his contract, wanted to release his few head of cattle but Freek refused before compensation was paid for some damage Booy had allegedly caused. Booy appealed to the young deputy landdrost of Graaff-Reinet, Andries Stockenstrom, who sent the field cornet of Baviaansrivier, Philippus Opperman, to intervene – which he did with little success, being somewhat nervous of the Bezuidenhouts.

Booy went away to join the Cape Regiment but the following year, accompanied by one of his countrymen called Dikkop, returned to claim his cattle, whereupon he was beaten up by Freek, or so he claimed. Subsequently, to the mounting fury of Stockenstrom, Freek gave a series of excuses to avoid answering the various summonses sent in his direction. Things took a trickier turn for Freek in May 1814. The unhappy Booy was caught by Dikkop in the arms of the latter's wife and, in the ensuing scuffle, Booy tried to stab Dikkop with an assegai while his rival more successfully shot him dead.

Dikkop hid the body and went to work for Freek, who refrained from following up on the disappearance of Booy, his erstwhile employee. But Dikkop's wife told the story to a domestic servant of Gerrit Bezuidenhout who naturally told Gerrit's wife who told Gerrit who told Opperman, who was once again not anxious to follow up on it. Because he thought he might (unjustly) be accused of the murder of Booy, Freek henceforth had even greater motive for avoiding any summonses.

Finally, in early October, the court messenger of Cradock, J.J. Schindehutte, was deputed to bring Freek before the circuit court in Graaff-Reinet. However, Schindehutte heard that Freek had threatened to string up the next court messenger who crossed his path and shied off. Opperman then reported that Freek was on crutches and unable to travel, only to change his tune soon after and give him a clean bill of health. The court, losing patience, found Freek guilty of contempt of court and sentenced him to a month in jail. The courageous Opperman hastily declined to make the arrest. When the field cornet at Buffelshoek at first balked at the task and then fell off his horse after finally agreeing to undertake it, the messenger J. Londt went to Captain Andrews, commander of the military post at Vanaardtspos. He, in turn, passed the job on to Lieutenant Frans Rossouw in charge of the military post at Rooiwal, closer to Baviaansrivier. Andrews ordered Rossouw to go with Londt and a trooper called Gerrit Lemke to assist in the arrest of Freek Bezuidenhout. They were voluntarily accompanied by Ensign William McKay of the Krugerspos garrison who joined his four pandours to the eight of Rossouw. So the posse of four whites and twelve pandours headed for the barren and stony Baviaansrivier valley.

Freek Bezuidenhout was already aware that something was afoot. He ordered the two youngsters on his farm, Jacob Erasmus and Frans Labuscagne, to keep watch, so he was not taken by surprise when the arresting party first appeared at nine o'clock on the morning of 10 October. His 'wife' thrust a gun into his hand and he, his servant Hans (who may have been his son) and Erasmus ran to a nearby cave, shouting to the soldiers not to come any closer. When the soldiers did not answer he and Hans started shooting.

Roussouw gave his men the order to fix bayonets but specifically told them not to fire. The cave was an excellent defensive spot with an entrance so narrow that only one attacker at a time could negotiate it, always aware of the expert marksman lying in wait at the other end. Two of the pandours climbed up above the cave and called down to him to surrender. He swore back at them – pointing out that if he did so they would shoot or hang him. For an hour Londt and Rossouw tried to persuade him to change his mind, the latter nervous that the gunfire might attract the attention of Freek's neighbours.

The pandour sergeant, called Joseph, with consummate courage crept up to the

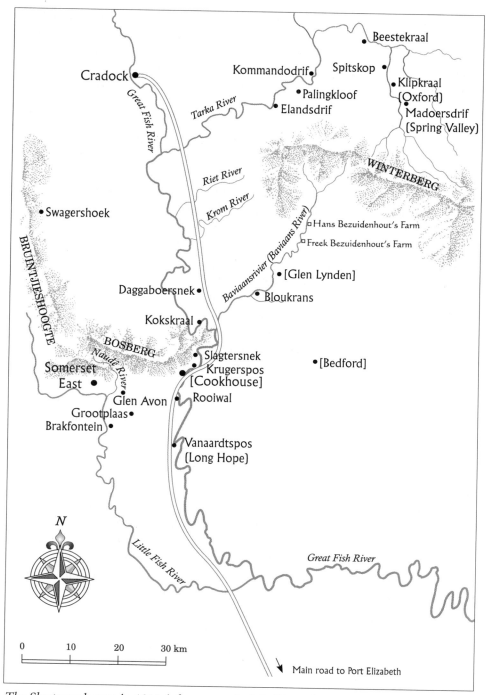

The Slagtersnek area in 1815 (after a map in J.A. Heese Slagtersnek en sy Mense*)*

cave entrance, flattening himself against the great rock. He, too, tried to reason with Freek who replied that they would have to kill him to capture him and from now on he was not going to waste his breath talking. Hans later stated that Freek was then knocked down by a stone thrown from above and then shot as he staggered to his feet. Lemke said Freek shot at Rossouw. Londt was sure that Freek was about to fire when a pandour shouted a warning to Joseph. Whatever the truth it is certain that Joseph and a colleague, David, shot Freek dead. (Freek's brother Gerrit afterwards was reported to have said that Freek fired the first shot.)

Hans and Erasmus gave themselves up and after a hasty examination of the cave the soldiers withdrew. Within a mile they were overtaken by Gerrit Bezuidenhout and four of his sons, curious to know if there had been a Xhosa incursion. But Rossouw ordered his men to keep quiet and got them safely back to their posts. Londt and Lemke continued on their way, with their two prisoners, to Graaff-Reinet. As they passed Naudérivier, under the Bosberg, they made a minor decision which had major consequences. They persuaded a farmer there – Hendrik Frederik 'Kasteel' Prinslo – to help them guard the prisoners on their journey.

Two days later a group of some 30 people attended the funeral of Freek Bezuidenhout. Amongst those gathered there were Groot Willem Prinslo and Piet Prinslo (a.k.a. Piet Kafferland), Wijnand (the eldest brother), Gerrit and Hans Bezuidenhout, Gert Krugel, Lucas van Vuuren, Frans Labuscagne, Cornelis Faber (Freek's brother-in-law), Louis Fourie, Frans Smit, Christiaan Dreijer, old Van der Sand, the school teacher, and others.

Hans Bezuidenhout spoke with bitterness about his brother's death, and swore vengeance – even if it took ten years – on Opperman, Stockenstrom and Rossouw, all of whom he held accountable. He asked the bystanders how they were disposed, and when some were silent, he swore an oath that he would spare neither woman nor child. He rounded on Lucas van Vuuren and Frans Labuscagne in particular, pointing at them and saying that they were 'the cause of it' (the reason for this is obscure but it was probably because they declined to support any acts of revenge). He said they 'were even worse than wolf turd'. Subsequently Freek's brothers made flailing attempts to send memorials of protest and complaint to the authorities.

By the end of October, however, Hans had not succeeded in stirring up much interest in his personal vendetta and found himself on the farm Elandsdrif near Cradock preparing to trek away in disgust. Things changed radically with the arrival there of Hendrik Frederik Prinslo, popularly known as Kasteel. Prinslo was the son of Marthinus Prinslo who had been condemned to death in 1799 (the sentence was commuted and the prisoner spent three years in the Cape Town Castle, which is how his son acquired his nickname to distinguish him from another burgher of the same name).

Oupa Prinslo had come to Bruintjieshoogte as early as 1770 and members of the family were always, as one historian of the rebellion, J.A. Heese, writes, the 'voorbokke' (the lead goats) whenever the frontier farmers found themselves in opposition to the authorities. Kasteel himself seemed to be pursuing the grievances of his father. When, therefore, he rode the 80 kilometres to Elandsdrif he injected a new element into the situation – a social dimension, however inchoate.

Hans and Kasteel found their interests coalesced and they formulated a plan. It involved an alliance with the Xhosa. The pair rode over to the farm of Stephanus Bothma where they invoked the help of three of Chief Jalousa's men to sound out Chief Ngqika about a possible attack by the Xhosa against the Rooiwal and Krugerspos military posts and even a more ambitious invasion. They went further: Hans instructed Cornelis Faber, accompanied by Adriaan Engelbrecht and Frans Marais, to go to Ngqika to solicit his backing. Faber afterwards claimed he tried to get out of the trip because 'he was troubled with the piles' but Hans 'forced' him to go.

Kasteel went back to Bruintjieshoogte to drum up support while Hans confided their plans to the owner of Elandsdrif, Diederik Muller. The latter, worried by this information, reported it to his field cornet, Stephanus van Wijk, of the Tarka district, who confronted Hans over the matter and urged him to petition the government over his grievances.

As one of the sheepish footsoldiers of the rebellion Marais' story is not without interest. Known as the 'Frenchman' (perhaps because of his first name), he was actually a Hungarian, 29 years of age, who had come to the Cape as an artilleryman for the Dutch. Together with two Poles he ran away from the battle of Blaauwberg to Swellendam and made his way by wagon into the interior, living on several farms along the way. He tried, without success, to get a pass from Stockenstrom to be in the area but instead got a job as a police rider for the deputy landdrost, Van der Graaff, in Cradock. He soon ran up debts and, in his words, 'innocently got a flogging' from his employers. He also spent some time in the drostdy in Uitenhage, having been arrested by Opperman for not having a pass. He had even lived for six months beyond the Kei River with the Paramount Chief Hintsa and he had been three years eking out a living as a shoemaker when Hans lent on him to go with Faber to Ngqika. He was nervous about the mission and before they reached Xhosa territory he expressed his misgivings to Engelbrecht. The small party went to the place of Chief Jalousa, who was not at home. The next day they rode to Ngqika's.

At that chief's Great Place, a 'Gonaqua Hottentot', Hendrik Nouka, was quickly summoned to act as interpreter. Nouka found there Faber and 'a young, tall person' (Engelbrecht) and 'a European fellow, short and broad' (Marais). He later testified at the trial that Faber said that he did not speak in the name of Prinslo and Bezuidenhout but in that of 'the people'. Faber told Ngqika that the whole of the Baviaansrivier were united as well as all the people through Graaff-Reinet to the Cape and there were 'six hundred Hollanders who were also ready'. Their object was 'to drive the English into the water again' and if the Xhosa did not help them, the English would attack them first, and then the Xhosa. Faber complained that the English made the farms too small and when the people said to them that 'the cattle eat one another dead' they merely told them 'to sell the cattle and make them fewer'. Faber added that formerly the English 'apprehended us, but now they shoot us'. When Ngqika asked him who, he replied Frederik Bezuidenhout. He even went so far as to suggest that the English would come with a parcel of soldiers to speak to Ngqika and then would treacherously shoot him.

Ngqika was canny, promising nothing. He said he would inform his uncle, 'Slambie' (Ndlambe), who, it was known, 'wept every day for the Zuurveld' which he had lost. No doubt he wanted to consult with Paramount Chief Hintsa, too. According to Nouka, Faber promised the Xhosa that they would get back the Zuurveld as well as the beads, brass, iron, pots, guns, powder and shot of the soldiers and the burghers who refused to join the uprising. In return the farmers would extend their territory up to the Koonap River. Faber spoke most, though the short broad man was very lively and talked a great deal whereas the tall young man was silent, and evidently there against his will.

Faber and his masters knew well that this embassy to the Xhosa constituted High Treason and that the risk they were taking was enormous.

On 5 November a visiting pastor from George (T.J. Herold) held a service which was attended by many people from Bruintjieshoogte, Swagershoek and Baviaansrivier. There was undoubtedly much secret and heated discussion. The field cornet of Baviaansrivier, Opperman, got wind of this, knew that Hans had long had 'an evil eye on him' and fled with his family to Graaff-Reinet, passing authority on to his deputy, Willem Krugel.

One of the most vociferous of Kasteel's supporters was Theunis de Klerk (of the farm Daggaboersnek) and the two of them arrived at Elandsdrif on 9 November. Faber was back from his mission, Stephanus Bothma was summoned by Hans Bezuidenhout, and Andries Meijer and Zacharias de Beer were also present. It was decided that Faber would return to Ngqika to pin down his participation and Bothma wrote a letter (probably dictated by Hans and Kasteel) to a relative of Kasteel's which Kasteel himself signed. By signing this 9 November letter, Kasteel to all intents and purposes signed his own death sentence, so it is worth quoting in full (though it loses a bit of its coherence in the contemporary translation).

Dear and much esteemed Cousin, Jacobus Krugel, I wish you the most necessary for Soul and Body. Cousin, I write to you in the Name of the Burghers of the whole of Bruintjes Hoogte, Zuurveld and Tarka to represent the Business to your District, and especially the Field Cornet, Van der Walt, that we have unanimously resolved, according to our Oath, which we took to our Mother Country, to remain as Protectors to remove the God forgotten Tyrants and Villains, as every one, let him be who he may, is convinced with God, how shocking, and how God forgotten it goes with our Country, which we took an Oath for, for every one is convinced, whether or not they shall be present at the appointed date; and to you I trust the Business to bring it under the people's eyes as speedily as possible, whether they will or not, and I send you the letter in the hands of the Burgher,

Christiaan Muller, and request an answer with the Bearer what the people say; the consequences speak for themselves, I trust to you to bring it under the people's eyes.

And this letter I recommend in your hands to burn; you see my great confidence in you, the letter serves you all. I therefore hope you will burn it directly you bring it under the people's eyes verbally.

Now I trust in you, and am, with esteem and greetings to you, your Cousin.

(Signed) Hendrik Fredrik Prinslo

The 9th November, 1815.

This letter was given to Johannes Muller to deliver. It burnt a hole in his hands and his elder brother advised him to take it straight to Field Cornet Van Wijk. This officer raced to Cradock to show it to Deputy Landdrost Van der Graaff who immediately warned Stockenstrom in Graaff-Reinet. Van der Graaff also sent instructions to Acting Field Cornet Willem Krugel who passed them on to Captain Andrews at Vanaardtspos and Ensign McKay at Krugerspos. Andrews was provided with a copy of Kasteel's letter and orders to arrest him. On 13 November this was done. Andrews was careful this time to send white troops – Sergeant Cooper and a detachment of dragoons – since one of the sores that had so festered was that Freek Bezuidenhout was killed by pandours. Kasteel did not resist – he was asleep when captured. He was taken to Vanaardtspos.

The authorities acted with speed. Alerted to the trouble, the field commandant of Uitenhage, Willem Nel, hastened to secure the transportation of Kasteel to Vanaardtspos (in the nick of time, because Theunis de Klerk organised a belated rescue party). Stockenstrom spent 40 hours in the saddle riding to Cradock and 'straining every nerve to guard the Fire which was kindling'. Major George Fraser rode from Grahamstown with some dragoons to Vanaardtspos arriving at noon on the 14th, but having only 36 men in all, felt there was little he could do to relieve Rossouw at Rooiwal and McKay at Krugerspos since he believed there were '200 mounted Farmers in arms' against him (there were probably never more than 65, if that). The landdrost of Uitenhage, Jacob Cuyler, also hastened to Vanaardtspos to take charge.

At a confused meeting at Daniel Erasmus' farm in the Baviaansrivier district on the evening of the 13th, Acting Field Cornet Willem Krugel asked a number of the burghers if they were loyal to the government. At least a few, headed by Theunis de Klerk, demurred. Krugel sent round official messengers to the men under his new command to meet him next day, armed with four days' provisions, at Slagtersnek (probably on or close to his own farm of Kokskraal), citing the imminent threat of Xhosa invasion. (Hans Bezuidenhout was at this time threatening a 'War of the Bloodpit' by which he seems to have meant that, through the Xhosa, much blood would be shed.)

In between these two meetings Krugel, either intimidated by Hans Bezuidenhout and De Klerk or because of his blood tie with Kasteel (they were cousins), changed sides so that the commando members who had been summoned on government business suddenly found themselves in a rebellious situation! But his defection did give the rebellion new impetus (Louis Tregardt, who was to win fame as a prominent Voortrekker leader, managed to avoid recruitment by perspicaciously absenting himself from his home).

The band of rebels now moved southwards to Vanaardtspos to demand the release of Kasteel. On the way they passed Rooiwal and Hans Bezuidenhout wanted to attack it, to revenge himself on Lieutenant Rossouw with, in his own words, 'Devil's violence'. He was persuaded, only with great difficulty, to leave this for later.

The rebels stood a kilometre or two off from Vanaardtspos and three separate messengers were sent in, to demand the release of the prisoner – 'in the name of all'. The second of these, Nicolaas Prinslo, took the opportunity to do a little reconnaissance – he noted that there were four officers in the house and estimated that there were no more than 20 soldiers, but that they were well concealed behind a large dunghill which afforded them effective protection. The rebels eventually asked to speak with Willem Nel, the field commandant.

Despite the offer of relatively safe passage Fraser was reluctant to allow Nel to expose himself to danger. Feeling ran against Nel because he had ensured Kasteel's imprisonment. But, even though he was having trouble with his eyes, Nel courageously rode out to negotiate with the rebels. He got nowhere. On the contrary, there was a tense moment when he was about to leave. Theunis de Klerk and Hans Bezuidenhout took hold of his bridle and compelled him to dismount. De Klerk threateningly said to him something along the lines of 'I shall shoot you or you shall shoot me.' When Hans said that he would revenge the blood of his brother Nel suggested that he must leave revenge to God and, if he believed his brother innocently killed, send a written complaint to Major Fraser. 'That,' Hans replied, 'you may wipe your backside with.'

Only the intervention of Piet Kafferland defused the incident and freed Nel. After this Fraser sent a poor teacher called Touchon with a copy of Kasteel's 9 November letter to be read to the rebels. When they heard it some of them for the first time began to understand the gravity of their situation. But De Klerk and Hans told them that Kasteel had not written the letter (technically Stephanus Bothma had written it) and made them gather in a circle where Krugel recited an oath of loyalty amongst themselves. Some of them said 'Yes', others took off their hats in silent acquiescence. If their evidence at the subsequent trial is to be believed almost all of them were not in the ring but looking the other way, tending to horses, or whistling to the wind.

Frustrated in their attempt to breach the fort and free their Castle the rebels retreated across the Fish via Jager's Ford and rested themselves and their horses on Louw Erasmus's farm at Kwaggahoeknek opposite Lt. Rossouw's post at Rooiwal. Despite the oath it is likely that Krugel, having heard the contents of the 9 November letter and weighed its implications, began to waver and regret his somersault. Nevertheless, A.C. Bothma was sent with a letter to Swagershoek to find fresh recruits. Few came, especially after hearing that the Xhosa had been invited in; in fact, the rebels began to leak deserters, who would slip away when they felt that they could escape Hans Bezuidenhout's ever-present threat to put 'a ball through the head'.

Lieutenant-Colonel Jacob Cuyler, commander of Fort Frederick at Algoa Bay, landdrost of Uitenhage and head of the military forces on the eastern frontier, was actually an American of Dutch ancestry and born in Albany, New York (he named the area in the eastern Cape after his birthplace). He was a man who did not brook contradiction and his temper was greatly feared. When he finally arrived at Vanaardtspos at sunset on the 16th he considered declaring martial law (as he had the right to do) but refrained from going that far and instead tried to open negotiations with the rebels though these were firmly snubbed by Bezuidenhout and De Klerk. Because they hoped that Van der Graaff might intervene more favourably they moved to Slagtersnek, closer to Cradock, while Krugel used the more neutral Groot Willem Prinslo as an envoy to Cuyler. The latter, however, felt he could only accept unconditional surrender.

So, on the morning of 18 November, Cuyler approached the 'Rioters' at Esterhuispoort, who were assembled on the stony hillock near the place called Slagtersnek. His force consisted of 40 dragoons and of 30 trusted burghers under the command of Willem Nel. When they came in view Bezuidenhout angrily said to Groot Willem, 'My God, nephew Willem, have you betrayed us again?' He ordered the men to retire up Slagtersnek hill to reduce the effectiveness of the dragoons.

Cornelis Faber seems to have been one of those people with an unfortunate sense of timing. On his second mission to the Xhosa in the middle of his exhortation to Ngqika to join the rebels Xhosa intelligence brought in the news of Kasteel's arrest and, therefore, the exposing of the whole plot. This put an end to any chance of Ngqika participating in it. Now, as the rebels were gathered facing the advancing government forces, Faber rode in with the bleak message of rejection from the Xhosa. It made Bezuidenhout very angry but undoubtedly undermined the confidence of most of the others.

Cuyler, sensing that the rebels were uncertain and divided, made several attempts to persuade them to give up. Hendrik Lange twice approached them: on the first occasion, Krugel offered surrender in return for a general pardon; on the second, Lange took with him a list of the leaders with the offer that if they gave themselves up the rest would be allowed to return home but though Krugel wept continually, feeling his guilt, De Klerk stated that *he* 'would be damn'd before he surrendered'.

So Cuyler arranged his order of battle: in the centre he placed the dragoons; some of the burghers were positioned on the right wing under Nel, the rest on the left under Field Cornet Bekker. Towards the top of the hill the rebels took up a position in the form of a crescent, about two paces between each of them, Krugel on the right wing, Bezuidenhout amongst other leaders menacingly on the left, threatening anyone who rode away that 'he would shoot him so that his brains would fly out of his throat', anatomically difficult perhaps but the gist was obvious.

The situation was a dangerous standoff. Cuyler called to the rebels to come down and the innocent would be pardoned; the rebels for their part waved their hats and cried out to the loyalist burghers to separate themselves from the soldiers.

After a threatening advance uphill by the government force Lange jumped off his horse, which could no longer carry him up because of the steepness of the ascent, and ran amongst the rebels saying that they might shoot him but if that were to happen 'all of them should be pursued to the very end of Africa'.

At this, some of those on the right wing – most of them young men – began 'to waver and to weep', and Krugel surrendered himself with tears in his eyes, saying that he felt he was guilty and awaited in God's name his equitable punishment. Then Lange placed his hand on Groot Willem's arm and called him 'Brother'. Groot Willem allowed himself to be led down, turning to the remaining rebels and saying, 'Brothers, you see that I am not the first, as some have already surrendered before me.' Whereupon the unrepentant rebels, with Theunis de Klerk and Hans Bezuidenhout at their head, rode away at great speed in the direction of the Tarka. In the remaining light of the setting sun the prisoners were herded away to the military posts and ultimately to Uitenhage.

It was left to Major Fraser to track down the fugitives. He left Vanaardtspos on 24 November with a force of 100 pandours and 22 burghers under Willem Nel but had a serious fall from his horse and broke his arm. He had to relinquish his command to Lieutenant McInnes and Ensign McKay. They reached Daniel Erasmus' farm at Baviaansrivier and stayed there until the evening of the 26th when they marched away at two o'clock in the morning in the direction of Johannes Bezuidenhout's place.

Casting around over the next couple of days they came across two of the rebels – Abraham Bothma and Andries Meijer – who gave them information about Bezuidenhout's movements. Hans planned to move down from the Winterberg with his wife, Cornelis Faber and Stephanus Bothma and four wagons, and make his escape into Xhosa territory via Madoersdrif. So, at daybreak on the 29th, McInnes set his ambush in the kloof (made by one of the tributaries of the Tarka River) through which the Bezuidenhout party must pass. McInnes proceeded down-river himself while Nel, his burghers, a sergeant and 18 mounted men of the Cape Regiment were stationed higher up the river.

After noon four wagons with cattle, sheep, goats and horses entered the kloof and unyoked, oblivious of the soldiers who encircled them in concealed positions. Two men, one armed and on horseback, the other unarmed and on foot (Faber and Stephanus Bothma), moved towards the place where McInnes and McKay lay. A party of six men were situated on the road in front of the rebels and McInnes, when he saw they would be discovered, ordered them to stand up. At the same time he revealed himself and called on the pair of rebels to stand.

Faber immediately turned on his horse and galloped off. McKay fired a shot above his head but he did not stop so the soldiers sent five or six shots after him. Thereupon Faber dismounted, kneeled down and presented his gun with a view to firing. Before he could do so he was shot in the left shoulder.

Bothma meanwhile had hot-footed it up the mountain, spat at by the guns of several soldiers. He found a hole and crept into it, where he was found and taken, along with Faber. McInnes then closed in on the wagons.

On the first shots being fired at her brother Martha Bezuidenhout (née Faber) left her wagon to see where they came from. She saw her husband mount his horse, preparing to gallop off. But when she asked him if he would leave her and her children to be killed, he dismounted and stood by her behind the wagon. He then spoke to the soldiers in front of him but when one of them prepared to fire she shouted a warning at her husband and pushed him out of the way. Bezuidenhout then fired, hitting one of the soldiers. He then stood behind the wheel of the wagon, reloading. A shot hit him in the arm breaking the bone above and below the elbow. He ran to his wife in order to hide himself and she called to the soldiers to lay hold of him but, as she covered him, she herself was wounded (there was later some query over whether she herself had fired a gun). Hans then hastened away from her but was shot down. Martha ran to him and lifted him up, when she was again wounded.

Hans said to his 14-year-old son, 'Go to them and they will not do you any harm'. The boy did so, but was wounded in the leg in the process. Ensign Mckay, coming up from the river, found Bezuidenhut lying on his back between the two rear wheels of a wagon. He seemed to be praying and appeared to be severely wounded in several places, judging by the amount of blood on the ground. McKay proposed to Martha Bezuidenhout that she bind up her husband's wounds and wash him but she refused, because of her own wounds.

In the wagons a cache of ten muskets and rifles (the largest loaded with 'slugs') was found, as well as three pigs of lead, and some horns and knapsacks filled with about 40 or 50 pounds of gunpowder. McKay went over to the wounded pandour and found him peppered with slugs from his left breast to his left thigh and his left arm broken. He died a couple of hours afterwards.

Hans Bezuidenhout did not die immediately; he passed away that evening as the sun was setting.

The pandour was buried on the farm Oxford, Bezuidenhout on Spitskop (afterwards Rocklands).

The remaining rebels were systematically gathered in or surrendered. Theunis de Klerk's first son was born four days after Slagtersnek and De Klerk gave himself up not long after.

The trial began on 15 December in Uitenhage, presided over by the Circuit Court judges P. Diemel and W. Hiddingh, with Cuyler prosecuting. The record – all 979 pages of it – makes fascinating, and moving, reading. The 47 accused did not have legal counsel and only asked a few desultory questions. But almost everyone involved was questioned. Not surprisingly most attributed their involvement (or non-involvement) to the threats of the fearsome Hans Bezuidenhout.

For instance, one of the unfortunates caught up in the turmoil and put on trial was Pieter Prinslo. He was a man of 50 years, residing in the Nieuwveld in the Graaff-Reinet magistracy. He had presented 'a memorial' to the government seeking permission to cross the Fish to look for 'herbs and domestic remedies' but this was not granted 'as one was not allowed to go beyond the boundaries'. Since all his herbs were used up he went to the regions of Bruintjieshoogte and Baviaansrivier to collect supplies from the mountains and forests. These he would administer for complaints 'which it had not been possible to cure' and from which he earned a subsistence. Amongst those he cured was the wife of Piet Erasmus whom the doctors could not help. When asked what these people gave for his ministrations he answered that 'they give what they like'.

That is how he came to be at Krugel's farm when Hans Bezuidenhout said he should go with him. When Prinslo asked, 'Where to, Cousin Hans?', he was given the response, 'To fight our country quite free.' Prinslo balked at this, saying he must remain with the sick woman (Mrs Erasmus) whereupon he was ordered to comply or expect the consequences, since the Xhosa 'had already come'. Hans told him he would bring the Xhosa in 'for as the English murderers had made use of Hottentots to shoot his brother' so he would make use of Xhosa murderers 'to shoot the English'.

So Prinslo was, in his version, coerced into accompanying the expedition to Captain Andrews' post. To the question, 'was he armed?' he replied, 'Yes, when I was in it, I was obliged but I had only three balls with me, and they were too large for my gun, and I purposely did not cut them smaller, because I would not do any harm.' He said that at the gathering near Vanaardtspos the rebel leaders 'spoke boldly' and added that 'I dare say they came there to do harm, but here was not any mischief done'. He maintained that he personally sat down with Nel and told him he would not be involved in the troubles. He could not recollect why Nel met the rebels 'for I was a forced man, in consequence of which my heart was sore'. But he did remember hearing Bezuidenhout say to Nel, 'You get your 300 rix-dollars a year to betray the burghers.' He denied taking Krugel's oath.

He also made the excuse that he could not slip away from Louw Erasmus's farm because his wagon was too heavily laden with timber to make coffins. He in the end admitted that he had 'been among the rebellious gang' but asserted that he was 'compelled' and prayed the court 'for mercy for a poor Sinner'. (He was sentenced to a year's hard labour on Robben Island.)

Most pitiful of all was the 16-year-old Adriaan Labuscagne who, when asked why he did not initially come down from Slagtersnek hill when Krugel surrendered, replied, 'I was fearful, and I am a young man who does not yet know what a Government is, as I was never near one.' (He was fined 50 rix-dollars as punishment.)

Despite the self-serving statements of most of the defendants, a clear picture of the whole episode does seem to emerge and justice, according to the times, was seen to be done though no summary of the events, motivations and circumstances can do them full justice.

What also becomes apparent – difficult perhaps to understand in this day and age with its sentimental aversion to capital punishment – is that hanging as a likely end to 'armed rebellion' and 'high treason' was accepted in matter-of-fact fashion.

Sentences were passed on 22 January 1816. A few were released, the majority were given a variety of punishments including banishment from the whole area. Six – De Klerk, Kasteel Prinslo, Faber, Stephanus Bothma, Abraham Bothma and Krugel – were condemned to death.

Some were rewarded for their part in the affair: Nel, branded a traitor by Bezuidenhout, was given a grant, to be made in perpetual quit rent, at a nominal rent, of his loan farm, Brakfontein; and Nouka was sent two heifers, and some plated buttons, tinder boxes and knives for his services. Ngqika was sent '11 fancy pictures' (one for each of his wives and one for himself), six tinder boxes, 16 knives, and three dozen handsome buttons, as well as 'some wholesome language', from Cuyler, though the landdrost was well aware 'that will as usual have little effect'.

Cuyler made a special appeal to Lord Charles Somerset on behalf of Krugel. He was, said Cuyler, 'a good, mild man who, from want of resolution to withstand the arts of designing men, has suffered himself to be led into the scrape'. Moreover, he had distinguished himself in recent fighting with the Xhosa and, as one of the first group to surrender at Slagtersnek, was a key to the ultimately relatively peaceful conclusion of that episode since 'their separating from the party at that instant broke the neck of the plot, and in all probability spared the effusion of blood, which at the very moment previous to their surrendering appeared unavoidable'. Krugel's sentence was commuted but he, with others like Frans Marais, was ordered to witness the executions and be publicly exhibited at the event with a rope around their necks to help concentrate their minds in future.

So, the stage was set for what turned out to be one of the more macabre events in South Africa's history. On 9 March five people of the Nek were sent to be hanged for the Nek by the Nek. But not at the Nek.

In fact, for subsequent propagandists, their martyrdom should not have been known by the name of Slagtersnek at all, since that was, in fact, a place of capitulation. But it does have a symbolically gruesome ring to it (though the name, in fact, preceded the rebellion and simply recorded nothing more sinister than the place where cattle were sometimes traded and slaughtered).

The site chosen for the executions was the small rise close to Vanaardtspos where the rebels had gathered round Krugel, Bezuidenhout and De Klerk to swear their oath of mutual loyalty. The hangman from George, together with his black assistant, was roped in to perform the ceremony with Revd. Herold in attendance to minister to the souls of the condemned. Friends and relatives of these unfortunates seem to have also attended in numbers and 300 military personnel ensured there was no trouble.

'The melancholy finish of the transaction was attended with every precaution,' Cuyler wrote to the Colonial Secretary in Cape Town, 'but an occurrence took place which made the scene more horrid and distressing.' This is an understatement.

The executioner, under the impression that he had a business relationship only with Hendrik Kasteel Prinslo, brought only one rope. Cuyler found that he could not buy any in Uitenhage so he was obliged to provide some which were lying in the Government store which 'although of sound appearance proved rotten'.

At the appointed time and place, Rev. Herold began with a prayer for the condemned, who requested to sing a hymn jointly with their rebellious companions and friends. To Cuyler this 'was done in a most clear voice, and was extremely impressive'. Stephanus Bothma actually addressed his friends, advising them to be cautious of their behaviour, and to 'take an example of me'.

But when the fatal drop came only the special rope held and the four faulty ones (despite having been doubled) broke. The four survivors, sprawled in the dust and no doubt bewildered, got up (who the lucky or unlucky fifth one was I have not been able to find out). One of them attempted to leave the spot and rush towards the place where the officials stood. All four of them spoke, and some of the spectators ran to Cuyler, 'soliciting pardon for them,' as he wrote, 'fancying it was in my power to grant it'.

They were all rehanged (again I cannot discover whether this was done one at a time with the one reliable rope). Cuyler was, despite his reputation, not an inhumane man and he was not immune to the 'distressed countenances' of the spectators but the law had its logic and its ritual and Cuyler could assure his superior that 'the prisoners one and all died fully resigned to their fate'.

Subsequently, and particularly in the twentieth century, Slagtersnek became a controversial debating point in historiography and a rallying cry for many Afrikaner nationalists in politics. But to go into the details of that is not our brief here. There is one curious anecdote to be added, however. It is contained in a letter written by Charles Meredith Harte that resides in the museum at Somerset East. It is the story of 'The Beam'.

Harte was born in Pietermaritzburg in 1886 of Irish parents. Being of 'an independent nature and anxious to farm', he bought an 'individual share' in the farm Erin's Hope, which was a portion of Verkeerde Water farm on the banks of the Great Fish River, in 1914. In the following year two aged neighbours individually volunteered the information that, according to their parents (possibly witnesses of the original execution), permission had been granted by Cuyler for the removal of the crossbeam from the gallows so that it could be preserved.

It was let into the walls of a thick stone-walled building nearby (subsequently used as a pigsty according to local legend), on land that was to form part of Harte's farm, Long Hope. It was rough-hewn and unstained, the ends pointed, perhaps to fit into the already existing structure, with four holes drilled in it for plugging onto the uprights.

In 1916, for the centenary of Slagtersnek, 11 special passenger trains disgorged hundreds of people at nearby Long Hope station. In subsequent discussion with Dr D.F. Malan (a future prime minister of South Africa) it began to dawn on Harte that the beam might have some commercial value and, when the ostrich feather trade collapsed, he became even more interested.

Interest over the years in the beam continued but there were usually some worries concerning its authenticity. In 1938 one of the oxwagons commemorating the Great Trek centenary passed through Long Hope on its way to Pretoria and its leader offered Harte £50 for the beam. That was much less than Harte was prepared to consider so he diplomatically deflected the proposition. To other interested parties he suggested that unscrupulous speculators might want to hack it up and sell the bits as 'sacred' relics.

Then in December 1947 Dr J. Albert Coetzee, steeped in eastern Cape history, arrived wanting to see the 'Slagtersnek Beam'. He also did a bit of impromptu archaeology suggesting that hundreds of cobblestones which Harte had dug up from a pathway and dumped in the river were evidence that Harte's farm had been the site of Vanaardtspos. Coetzee asked Harte if he would sell the beam and at what price? Harte said £500 and Coetzee replied, 'Don't you think that is rather much?' So Harte said, 'What are your ideas?' Coetzee offered 300, so Harte said, 'I'll split the difference.'

Four men, including a Mr Hulme who taught at Gill College in Somerset East, eventually came to witness the loosening and removal of the beam by Jafta, one of Harte's men, and its loading onto the chassis of a lorry to which no body had been fitted. Payment was made in February.

In after times exaggerated stories were spread about how much had been paid for it. One rumour said that £3 000 had changed hands.

The beam ended up in the state archives in Pretoria.

There is a suggestion that it should return to the area. When I informally canvassed opinions sentiment was divided. Since the bicentenary is coming up soon a decision needs to be made, unless it is postponed for another hundred years while scars heal and tempers cool.

CHAPTER SEVEN

—•—

AMALINDE

The Battle of Amalinde has all the marks of Homer: love, abduction, incest, cruelty, political machinations, the meddling of the fickle gods, deception and the clash of armies. It also has the most curious and intriguing terrain I've ever stumbled across. That is why the battle site is my favourite.

In 1897 the following remarkable poem – one of the very earliest in English by a black South African – was published in a black newspaper.

NTSIKANA'S VISION

> What 'thing', Ntsikana, was't that prompted thee
> To preach to thy dark countrymen beneath your tree?
> What sacred vision did that mind enthral
> Whils't thou lay dormant in thy cattle kraal?
>
> Was it the sun, uprising in his pride,
> That struck with glittering sheen Hulushe's dappled side,
> By Chumie's laughing fountain hastening merrily,
> To meet strong Keisi's waters rolling to the sea?

A Vision? Yea! That presence once had shone
Upon the men of Tarshish, down from the heavenly throne,
And in the holy light of His mysterious Word
The proud Barbarian bows and worships God, the Lord!

Hark! 'Tis the sound of prayer, of savage melody —
Untutored voices raised to Him who sits on high;
Those hills and dales around fair Gwali's stream
Repeat again Ntsikana's sacred Hymn!

Wake, Gaika, wake! I see the gathering storm
By Debe's plains; Gcaleka's horse and Ndlambe's legions swarm;
Behold thy tribesmen scattered, thy warriors' doom is sealed —
The word of God rejected — by prophecy revealed.

The poem was written by Allan Kirkland Soga, himself a member of a most remarkable family. The poem, in itself and within itself, reverberates with meaning and significance. One way of getting a 'feel' for it is by scouring the country's second-hand book stores in search of the book *The South-Eastern Bantu* published in 1930 by A.K's brother, John Henderson Soga. Much of the background to the poem is clarified in the book and the family's intimate connection to the Battle of Amalinde and its consequences are indirectly laid bare.

In the eighteenth century, the Xhosa lived largely to the east of the Kei River, under the well-to-do and popular Chief Phalo. But there was rivalry between his two most important sons, Gcaleka, prince royal and heir to the Great House, and Rarabe, of the Right Hand House. Driven by a termagant mother, Gcaleka tried to usurp the throne before his appointed time. He added to his unpopularity by studying to become a witch doctor – something frowned upon amongst a people where the separation of state and religion was the norm. (Gcaleka one day disappeared in a great pool at the Ngxingxolo River for a long time, during which miracles happened to him and he eventually re-emerged fully qualified in his new profession.)

His grab for power was opposed by his brother Rarabe, supporting his father, and war ensued. Who exactly prevailed is in dispute but the consequences were momentous. The House of Phalo was split. Gcaleka stayed on east of the Kei and gave his name to the Great House of the Xhosa: Rarabe crossed the Kei and established himself at the Izeli (near present-day King William's Town) and in the beautiful, green misty Amatole mountains, to which his successors and his people became bonded, like children to a mother.

Despite the riven relationship, the Right Hand House of the Rarabe always acknowledged the paramountcy of the Great House of the Gcaleka.

When Rarabe was killed in an excursion into Tembuland in about 1787 (he is buried near Mgwali), his grandson and heir Ngqika was three years old. Ndlambe, a son of Rarabe and Ngqika's uncle, became regent. A shrewd and powerful man,

Ndlambe ruled for 20 years, always with one eye on sovereign power.

As Ngqika grew to manhood the rivalry between the two also grew. Ngqika was reckless and feckless. He fell incestuously in love with the beautiful Thuthula (daughter of Dibi and one of Ndlambe's wives) and abducted her. This incestuous liaison was a remote but direct cause of the Battle of Amalinde.

One of the more immediate causes of the war came in 1817 when Ngqika met the governor of the Cape, Lord Charles Somerset, and was forced to cede the territory between the Fish and Keiskamma rivers. The Xhosa felt that since Ngqika was not the paramount chief his was not the authority to alienate the land. Ndlambe duly exploited the resentment.

There were other factors – personalities – at work. Two men out of the ordinary arose to influence in the two camps. They were what Noel Mostert in his book *Frontiers* called 'charismatic millenarian prophets'. Both of them at a fairly early age came under the sway of the missionary, Johannes van der Kemp, stationed at Bethelsdorp near Port Elizabeth.

Nxele (otherwise known as Makanna) proclaimed himself the son of God and promised the resurrection of the dead on a certain day. Although this did not happen Nxele did offer to Ndlambe and his followers a way of thinking which provided them with some exploration of and resistance to the pressures being exerted by the encroaching whites. Ndlambe used the growing influence of Nxele in his power struggle with Ngqika.

Ngqika, in his turn, sought advice from another seer, Ntsikana. The latter had not only heard Van der Kemp preach at Ngqika's Great Place, but was also in contact with another missionary, Joseph Williams. One day Ntsikana arose and went to his cattle kraal where he found his favourite ox, Hulushe, surrounded by a strange and bright halo. The spirit of God ('the thing') entered the soul of Ntsikana. He began to preach his own syncretic version of Christianity as a messenger sent by 'the Chief of Heaven and Earth'. He also composed several hymns – the first literary compositions which can be ascribed to an individually named black South African. His 'Great Hymn' can to this day be heard ringing through the valleys of the Eastern Cape on a Sunday. Although he first offered his visionary services to Ndlambe, who rejected them since he was already served by Nxele, he found his spiritual home under the wing of Ngqika.

It was inevitable that the animosity between Ngqika and Ndlambe, the frog and the snake, would lead to open conflict. It came in 1818 as the result of a severe drought. In October, the lands of Ngqika, resident at the Keiskamma, were scorched and the rivers dry. Ngqika decided to visit his uncle, Ndlambe, who was living in the neighbourhood of the Buffalo and Peelton rivers. On the way he was surprised to find the fields of Ndlambe lush and green, with royal herds grazing in the charge of Hintsa, grandson of Gcaleka and soon-to-be paramount chief.

Ngqika went away and returned with his own herds which he settled on his uncle's pasturage. The herdsmen attacked each other and Hintsa was driven off. This insult to the royal household led them to ally themselves with Ndlambe. But the Gcaleka were repulsed by the followers of Ngqika near Hoho mountain (between Middeldrift and King William's Town).

Ngqika requested a meeting with Hintsa, assuring him of his safety. But when the younger man, Hintsa, appeared before him the older man harangued him: 'My elder brother's child, why have you come against me armed, in a matter that affects Ndlambe and myself alone? You, the representative of the Great House, making war upon me, your Right Hand! Why have you done this? Here I hand you four assegais (*intshuntshe*); they are the heritage of the Xhosa tribe. Let them be a symbol of peace between us. Let this matter be ended. It is late, already dark, we will speak further in the morning.' Fearing for his life, Hintsa fled during the night.

Separately, the forces of Ndlambe and the Gcaleka were weaker than Ngqika's; allied, they were likely to prove superior. Injured pride was then the immediate cause of the fiercest battle ever fought by the Xhosa amongst themselves.

Ndlambe was urged on to war by the bellicose Nxele but Ntsikana took up a very different stance. He sent a message to Ngqika: 'Abandon all thought of war with Ndlambe, otherwise you will bring down upon yourself fearful retribution, the nature of which I cannot describe, but I see the heads of the Gaika being eaten by ants.' When Ngqika informed him that the Ndlambe warriors had already seized cattle of his, Ntsikana replied: 'Do not attempt to recover them.' But the Ndlambe army was already on the march so Ntsikana gave his chief a final warning: 'If you insist upon meeting them, beware lest the enemy entice you; do not follow them up, because you will fall into a dangerous ambush.'

The Battle of Amalinde took place on the Debe Flats. The Gaika forces came from the north and west, from the headwaters and valley of the Tyumie River (the 'Chumie' of Soga's poem). They probably reached Debe Nek just before midday. On the plain to the east they encountered a force of Ndlambe warriors. These the Ngqika immediately charged and put to flight. But the fugitives were the young and inexperienced. They were a clever decoy.

The army of Ndlambe was under the command of his son, Mdushane, a man of outstanding ability, wisdom and foresight. It is more than likely that it was his tactics that were used in the battle. He concealed his main army which contained the seasoned warriors who were privileged to wear the blue crane feathers of courage and seniority. As the Ngqika charged past Mdushane threw these men against their flanks.

From a safe spot on a mountain close by (almost certainly at Debe Nek) Ngqika watched his outnumbered army charge and counter-charge and his son Maqoma lead his section right into the heart of the enemy. He saw his son wounded several times and come close to being captured. He saw his men falling like wheat falls before the reapers. And he saw the remnants of his devastated army retreat under intense pressure.

The battle lasted from midday until near darkness. In the twilight the Ndlambe warriors used what was left of the light to dispatch the wounded on the battlefield. But they did not leave it there. In the darkness they built bonfires and ruthlessly tracked down any survivors groaning, crying out or trying to play dead in the dark.

Ngqika fled into the secure fastnesses of the Amatole and immediately appealed to Somerset for help (which was forthcoming). Few kraals amongst the Ngqika had not lost a son, and the womenfolk lamented, many Rachels mourning for their children.

Ngqika's power was broken; so was he. Drink became his new enemy and he died in 1829. He was succeeded by his crippled and ineffectual son, Sandile. Ndlambe's power was at its zenith and he soon was to be a real threat to the Cape Colony. Mdushane, the victor of Amalinde, died in 1828 of syphilis. Maqoma survived longest, to become a real thorn in the side of the British.

The account of these events explains much of the meaning of A.K. Soga's poem. But there is further significance to the fact of the poem's existence and to its additional meaning.

One of the casualties of the Battle of Amalinde was a prominent councillor of Ngqika, Jotello. His son was Soga who also became a renowned warrior and chiefly councillor. Soga's son, Tiyo, grandson of Jotello, was one of the early students who studied at Lovedale, the Glasgow Missionary Society's school newly established in the Tyumie valley.

Tiyo Soga went to Scotland, became the first black South African to be ordained in an established church, married a Scottish woman (Janet Burnside), and returned to the eastern Cape in 1857 to build a church at Mgwali (near Stutterheim). One of his first sad tasks was to help pick up the pieces of Xhosa society after the catastrophic Cattle-Killing which was the outcome of the (false) prophecies of Mhlakaza and Nongqawuse. A.K. Soga published his poem in a newspaper edited by his brother, William. The brothers were not without their social and political ambitions, starting one of those neo-political organisations of the 1890s which ultimately flowed into a greater political union in 1912.

In the poem Soga situates his family geographically (Mgwali) and historically. He links it to the pioneering figure of Ntsikana and establishes it squarely in a tradition of a visionary, progressive, educated, Christian elite, those who have accepted the 'true' revelation through the word of God.

Having long known and read about the battle I was most anxious to experience more tangible connections with it. There are two sites which are particularly moving.

The grave of Ngqika is situated above the village of Burnshill and not far from the Fort Cox Agricultural College (in the grounds of which are located the almost totally disappeared ruins of the fort where later so much fierce fighting was to take place). Ngqika's grave is a modest one, made of red stone. (You might have to ask locals for directions on how to find it.) The surroundings are superb and the site is impressive, perhaps because it is so forlorn, which is a feeling the visitor only in part brings along.

The second site moved me even more perhaps because I had not known of its existence and stumbled across knowledge of it only by accident. It is very difficult to find so its location is worth describing in exact detail. If you drive north 11 kilometres out of Peddie on the R345 dirt road you will come to an unmarked turn-off to the left which leads to Hlosini School (this is the only easily identifiable landmark to ask for). About 1.8 kilometres beyond the school at a red-roofed white hut there is a turning to the left. Three kilometres along from there you will reach a round hill with a fine 360 degree panoramic view. I arrived there after a journey with adventures even Don Quixote would have been proud of.

On top of the hill is an iron cross placed on a plastered-over plinth embedded in a concrete block. It is very simple and unmarked but, having lived with Soga's poem for many years, what a privilege it was for me to find it! An unintended pilgrim I was, at the grave of Ntsikana.

But the place I most wanted to see was the battle site of Amalinde. The battle gets its name from the Xhosa word which refers to depressions in the ground, variously described as cup- or saucer-shaped. Finding these is slightly easier said than done but the experience proved to be much more startling than I expected.

The Debe Flats lie alongside the main road (the R63) between Alice and King William's Town. In fact I approached this road from the north on the R352 from Keiskammahoek. At the T-junction where this road intersects with the main road (near Dimbaza) I noticed to my right a green field with undulations like ripples of water in the wind. Were these the *amalinde*? I turned right towards Debe Nek, that neck formed by the last of the mountains leading down from the Amatole.

To the left of the main road, heading towards the Nek, were the same undulations but it was drizzling and the atmosphere, not quite eerie, was nevertheless elusive. It was late afternoon and I needed petrol and accommodation so I made for King William's Town. There I asked a garage attendant if he knew the word 'amalinde'. Yes, he said. As a child he had played hide and seek amongst them near East London. What did they look like, I asked. Like heaps, he said, like grave heaps.

So I headed back for Debe Nek looking for furrows and barrows. I still was not sure whether the undulations I had noticed were them. I drove up and down, tired, looking for inspiration. Opposite a derelict hotel at the Nek a group of women was standing in the rain, chatting. One of them was a nurse called Cikizwa Sauli. She confirmed that the undulations I was seeing all around were *amalinde*. She took me to meet the local headman, Solomon Nyingwashe, who, she said, knew about the battle.

The version of the battle he gave me and showed me cannot be verified but should not be discounted.

To get the best feel for the battle, approach Debe Nek from the direction of Alice. The hill to the left was probably the vantage point of Ngqika. Over the Nek you look along the main road with Dimbaza on the right. About three or four kilometres along the road and just to the right of it is Pirie station. Drive to the turn-off to Dimbaza Central, turn right and park 20 metres along. This is where Solomon Nyingwashe brought me.

From there, parallel to the main road, is a gentle rise to Pirie station, dotted with the cups, bowls, dents, furrows and humps of the *amalinde* (particularly on the right). At the top of the rise, he said, was deployed the band of decoys in columns with the animal skins they wore turned back to front. The Ngqika must have thought they were looking the other way. 'We'll have this small morsel,' they said to themselves. But the main body of the Ndlambe, hidden beneath their skins and shields, were lying on their stomachs in the *amalinde*. This was the genius of Mdushane – to lure his enemy onto terrain that he could take advantage of.

I returned next morning to the scene in the bright sunlight. Was it really credible that such a trick could work? Were the *amalinde* deep enough to hide an army? I was not at all sure. But then I thought perhaps the day was overcast, with rain, or drizzle, or mist … Perhaps it *was* possible.

The last thing I asked Solomon Nyingwashe was what caused the *amalinde*? Was it erosion of some sort? Or ancient ploughing? My questions were naive, his answer startling. Earthworms, he said. Giant earthworms. The biggest in the world.

The largest known species of giant earthworm is found in South Africa. In 1967, for instance, a specimen was found crossing the national road at Debe Nek which measured 3.35 metres in length and 6.40 metres when naturally extended. The wormcasts of *Microchaetus microchaetus* can cause depressions or humps up to a metre high (if there was long winter grass still around perhaps these might well have provided effective concealment). The 'Debe Hollows' as they are referred to are called '*kommetjies*' (small basins) in Afrikaans. Their Xhosa name '*amalinde*' is derived from '*indebe*' meaning a ladle. Hence the place name Debe Nek.

Ntsikana had warned Ngqika about ants: one wonders if he ever thought of earthworms?

CHAPTER EIGHT

GRAHAMSTOWN

Made confident by his success at the Battle of Amalinde, Ndlambe gathered his forces for the conquest of the Zuurveld.

In December 1818, a column comprising regulars from the 38th and 72nd Regiments combined with some mounted burghers, all under the command of Lieutenant-Colonel Thomas Brereton, had pushed across the Fish River to harry Ndlambe. They were there partly at the request of the humiliated Ngqika, hoping to regain power from his hated uncle.

Ndlambe had retreated into his impenetrable fastnesses but Brereton confiscated 23 000 head of cattle of which he turned over 9 000 to Ngqika. When Brereton withdrew Ndlambe followed, raiding the countryside, burning farms and killing farmers wherever he could. Two small patrols were ambushed on separate occasions and Ensign Hunt (of the Royal African Corps) was killed in one of the ambushes and Captain Gethin (of the 72nd Regiment) in the other. The inhabitants of Albany gathered in laagers at Rautenbach's Drift and Addo while Nlambe's warriors marauded along the Sundays River and in the Zuurberg.

In April 1819 his lieutenant Makanna (or Nxele) approached the fledgling settlement of Grahamstown, backed by between 6 000 and 10 000 warriors. He was also backed by a fierce and idiosyncratic ideology.

He preached that there were two gods. One was Malidipho, god of black people. The other was Thixo, god of white people. Malidipho was stronger than Thixo. Because they had murdered the son of Thixo the whites had been expelled from their own land and had sailed over the seas in search of a new world. Their destruction was Makanna's recurrent preoccupation.

The now-discredited Brereton had been replaced by Lieutenant-Colonel Thomas Willshire of the 38th Regiment, who had rushed to Algoa Bay by sea in the *Alacrity*. 'Tiger Tom' was to head a force of some 3 000 men, consisting of burgher mounted commandos, infantry of the 38th and 72nd, men of the Royal African Corps and the Cape Regiment and a small detachment of the Royal Artillery, which was to push Ndlambe's armies out of the Zuurveld and pursue him across the Great Fish River. The final plan was to get this considerable force into the field by the end of May.

This plan was forestalled by the decisiveness of Makanna. Not only had he managed to gather an army bigger than anything the Xhosa had ever raised before but he had also successfully managed to conceal its movements.

At this time Grahamstown was barely an apology for a town. It consisted of no more than 30 houses, mostly of the long low type with thatched roofs, straggling along what was to become High Street. In amongst them (roughly where the Cathedral now stands) was an officers' mess. The most substantial building was then called the Wit Rug Kamp (subsequently known as the Eastern Barracks, then – as a mental asylum – Fort England). But it was a considerable distance off to the east and only accessible by paths through bushy and rough terrain.

The town was thinly defended. The garrison consisted of only 45 men of the Light Infantry Company of the 38th Regiment, 39 men of the Colonial Troop, 135 men of the Royal African Corps and 82 Khoikhoi soldiers of the Cape Regiment. These were augmented by 32 civilian inhabitants of the town – who were just about all its able-bodied men.

As he approached Grahamstown, Makanna was supremely confident. Never before had such a large Xhosa army been assembled. He also knew that the town was not strongly defended because he had a spy at its very heart. Nquka (a.k.a. Nouka, Nootka, Nutka, Ngcuka or Hendrik Nootke) was Ngqika's interpreter and go-between with the colonial forces and had carte-blanche to move between the two groups (he had played a small but significant role in the Slagtersnek debacle). But he was also passing intelligence to the amaNdlambe.

Further, Nquka was deputed to play an even more active part in weakening the town's defences. He informed Willshire on 19 April that there was some sort of disturbance towards the east, so the colonel – despite his wariness of Nquka – sent a patrol of Light Infantry, 100 strong, to investigate. Makanna meanwhile had swung his army round to approach Grahamstown from a more north-easterly direction. Arrogantly, Makanna/Nxele sent a warrior into the town on 21 April with a message to Willshire that 'we shall breakfast with you in the morning'. The colonel replied with his own message that everything would be ready for him when he came.

He does not seem to have taken Makanna's threat too seriously. The next morning (the 22nd) he himself rode out in the direction of Botha's Hill on an inspection control. At about 10 o'clock he noticed a large concentration of Xhosa far to the east, near Governor's Kop. He hastily galloped back to Grahamstown, narrowly escaping being cut off from its relative safety. At the same time a Khoi herdsman, on the flats above the town, observed the same phenomenon and dashed to the Eastern Barracks to warn its commander Captain Trappes (of the 72nd Regiment).

The advance elements of the Xhosa army were now on the doorstep. In fact, about noon some even entered the house of one Potgieter where the midday meal was about to be served. The Potgieters and the other civilians fled down High Street and sought shelter in the officers' mess building.

Trappes hastily deployed his troops. Sixty men of the Royal African Corps, under Lieutenant Cartwright, were selected to defend the barracks. The rest of the force was drawn up in a line facing north and stretching from the barracks across to where the modern railway station now stands. They were arrayed behind a small stream called the Matyala and looked up towards a small distinctive conical hill (now called Mount Zion) and a long ridge running away from it to the right. Along this ridge the Xhosa army, up to 10 000 strong, a swathe of red from the ochre daubed on the warriors' bodies, was spreading itself into attack formation. (The conical koppie is often mistaken for Makanna's Kop – in fact, it is the ridge which correctly bears the name.)

It is possible to place oneself almost exactly in Makanna's position at the moment just prior to the battle being joined. A drive north-eastwards up Beaufort Street (it extends into Raglan Street) passes through the venerable Fingo Village and up the ridge (ultimately to join the Fort Beaufort road). At the top of the ridge there is a rough track off to the right leading across a makeshift soccer field and following a line of telephone poles to the end of the ridge.

This is where Makanna stood, contemptuously looking down on the tiny settlement. He looked over to the sites where the City Hall and the Cathedral now stand, which may or may not have been where one of Ndlambe's Great Places and cattle kraal respectively had once stood. In any case, the scene seemed to symbolise the whole of the land he wanted to reclaim.

He made his preparations carefully, in no hurry, it seems, to launch his attack. A tenth of his force he sent a few kilometres to the east to ward off any relief that might come from a commando stationed at Blaauw Krantz, or from the detachment of Light Infantry that had been sent into that area. The bulk of his army he divided into three divisions: the one on the left, which he led personally, was to swoop down on the Eastern Barracks (the long red roofs of Fort England mark the site nowadays, and there is a small section of the original wall that still survives there). The other two columns were to head for the rudimentary High

Street and the officers' mess, confident of brushing aside the thin line of troops drawn up in front of them.

It was a significant moment – one of those knife-edge moments of history. 'Greater South African battles were to come,' historian Noel Mostert has written, 'but Grahamstown was the most significant battle of the nineteenth century in South Africa.' Preparations to bring thousands of British settlers to Albany were well under way. A military reverse at Grahamstown might have brought about an equivalent reversal in this immigration policy. And the future of the country may have been very different.

Makanna's men would initially have the advantage of rushing down a fairly steep slope and crossing the small stream which was not much of an obstacle. Then a gentle slope led up to the thin British line. But that slope allowed Willshire, who had by now returned, to position his soldiers to fire disciplined volleys from their smooth-bore muskets while over their heads the artillery guns could send shrapnel shells into the massed ranks of Xhosa. But the Xhosa did not come on until Willshire provoked them. The signal to charge was given by some gunfire amongst the Xhosa. The two right-hand divisions under Mdushane and Kobe (son of Chungwa, the Gqunukhwebe chief killed in Colonel Graham's revenge attacks after the treacherous slaying of Anders Stockenstrom in 1811), flung themselves down the hill towards the defenders who held their fire until they got within 35 paces of the British. One of the defenders fell to a Xhosa bullet but the line held and the artillery cut 'streets' through the Xhosa ranks. In the face of the withering fire the Xhosa did not retreat but crouched in front of the redcoats waiting for the opportunity to throw their spears. Their commanders ordered them to break the shafts and turn them into stabbing assegais. Meanwhile Willshire had been strengthened by the arrival of Boesak, an early Khoikhoi convert of the missionary Van der Kemp, and 130 of his followers, many of them buffalo hunters and expert sharpshooters. The hunters picked out the most prominent chiefs and warriors and when the Xhosa renewed their charge they were decimated and pushed back across the stream, which ran with the blood of the dead and dying so that to this day it is known as Egazini, the place of blood.

'Many of the Khosa,' said one witness, 'leaped into the deep pools of water between this and the barracks, merely keeping their head above the surface, which they endeavoured to conceal by covering them with such grass and weeds as overhung the banks, and so perished.' This part of the fighting, desperate and bloody as it was, lasted less than an hour. The fighting at the Barracks was even more intense.

In an action that anticipated Rorke's Drift, the Xhosa succeeded in getting over the outside walls and crossing the parade ground to the hospital section. The defenders maintained a punishing fire but were critically short of gunpowder. At this point a woman wearing a shawl – Elizabeth Salt, wife of one of the soldiers – walked unharmed through the Xhosa ranks and brought from the village a concealed and precious bag of gunpowder. By 3.30 Makanna finally knew it was over and he called off the attack. Within the walls of the Barracks 102 bodies were counted later.

The British losses were light – three killed (including Captain Huntly of the Royal African Corps) and five wounded. Xhosa losses were heavy – many hundreds, perhaps well over a thousand; even two thousand, according to one estimate.

Three of Ndlambe's sons died in the battle. Nquka fled from Grahamstown but was captured and shot in dubious circumstances, Willshire having determined that he should be hanged. His place as interpreter was taken by Hermanus otherwise known as Xogomesh (the area around what is now Fort Brown was at the time called Hermanus Kraal), who, in his turn, will reappear as a prominent leader of a future uprising.

Willshire crossed the Fish River and harried the refugees of the battle relentlessly. Four months later Makanna handed himself in to the British. He was transported to Robben Island, amongst the earliest of the famous prisoners there. With other prisoners he effected a daring escape but the longboat they were in smashed on some rocks at Blaauwberg and he was drowned. His name lives on in legend and myth.

Ndlambe died in 1828, aged 90. His last years were spent quietly in the sunshine, as he said, like 'an earthworm creeping out of his hole'. The chiefdom passed to Mdushane, 'victor of the epic battle of Amalinde', as John Milton has pointed out. But the power of Ndlambe's chiefdom was never the same after 1819.

The following year Ngqika died. He had many wives. One of them was Thuthula, with him to the end. By that time, though, his main love was Cape Smoke. It was probably brandy, then, that killed this handsome and intelligent man.

So these epical heroes in a tapestried landscape passed away in somewhat undignified and anticlimactic fashion, not cut down in battle but impersonally sidelined by the imperial equivalent of the Connecticut Yankee at the court of King Arthur.

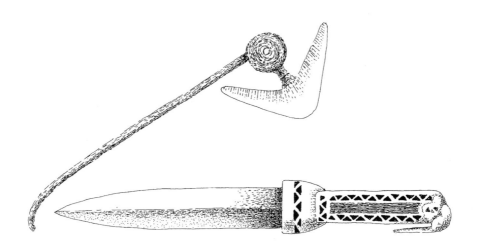

CHAPTER NINE

—◆—

LATTAKOO

Lattakoo is one of those evocative names of South African history. Was it one of its most significant battles?

By the beginning of June 1823, Robert Moffat of the London Missionary Society (LMS) had intelligence that a mysterious armed throng was cutting a swathe through the tribes to the north-east and was advancing on the Tlhaping under Mothibi and on his own mission station at Kuruman. The namelessness and facelessness of the marauding 'horde' simply added to the panic of the local inhabitants.

Moffat believed (probably mistakenly) that they were Mantatees, a section of the Batlokwa, a southern Sotho group who owed allegiance to their chieftainess MmaNtatisi.

On Friday, 6 June, Moffat set out in his wagon (he had no suitable horses) for Griqua Town, some 150 kilomotres to the south, to consult with Melvill, the British government agent there, and the Griqua chiefs and to plead for a commando 'to defend the country from the ravages of a foreign enemy'. He arrived there on Tuesday and was agreeably surprised to find George Thompson, a merchant from Cape Town, who had 'chosen to spend the winter in the interior'.

Melvill immediately put in motion measures for the country's protection; Andries Waterboer, the Griqua chief, rode off to Campbell to alert people there to the danger; and Thompson agreed to accompany Moffat back to Kuruman to encourage Mothibi and his people not to retreat.

For the return journey a horse, duly suitable, was found for Moffat and, much enjoying the company of his English friend, he arrived at Mothibi's town, Lattakoo, on the evening of the following day. On his father's death Mothibi had, according to custom, moved Lattakoo from its previous site – Old Lattakoo – some distance to its new site. (Mothibi himself had some years before removed to Kuruman.)

On Friday 13th, a great pitso took place on the edge of Kuruman. It began with much singing and mock fighting. Then the whole crowd sat down in the form of a crescent and Mothibi (in a white linen garment), Phethu (a young prince in an officer's coat) and eight or nine speakers discussed the coming war.

On the 16th Moffat accompanied Thompson as far as the Matlhwaring River since the latter wanted to see the old town of Lattakoo and to discover 'some more authentic and direct information' about the invaders. But Moffat's horses were not in good condition so he turned back but not before meeting people who had fled from Nacuning where an attack was expected at any moment.

Thompson was determined to see Old Lattakoo (some 40 miles north-east of Kuruman) and make contact with the advancing marauders. The perilous nature of this adventure adds a real edge to Thompson's account.

Accompanied by a man called Arend, he set out from Kuruman in the heavily dew-laden morning of 20 June. They crossed a perfectly level plain, covered with fine grass and 'bounded on all sides only by the horizon'. After several hours riding they descended 'a gentle eminence' to reach the spot where the old town of Lattakoo stood. When John Campbell visited it in 1813 it stretched three or four miles along the valley and had at least 1 500 households with, perhaps, a population of about 15 000.

As he approached the new town Thompson was pleasantly surprised by the extensive fields of millet giving signs of industry and prosperity, but the unusual stillness of the fields and the town itself put him immediately on the alert. He found the town, usually alive with 6 000 to 8 000 inhabitants, 'as solitary and silent as the most secluded wilderness'. Half-prepared food stood in pots in eerie silence. Hoping to attract anyone in hiding with the sound of a musket report he took aim at a large white vulture 'which sat perched like the genius of desolation upon a tall camel-thorn that shaded the residence of some chieftain' and 'brought it fluttering to the ground'. No living thing responded.

Arend was anxious to leave – their horses were weak and the insurgents must be close at hand. But Thompson felt that would thwart his whole purpose in coming, so they pushed on to the north-east. Just as they were about to venture to the river to refresh themselves and water the horses Arend cried out 'The Mantatees! The Mantatees! – we are surrounded!' They could see 'an immense

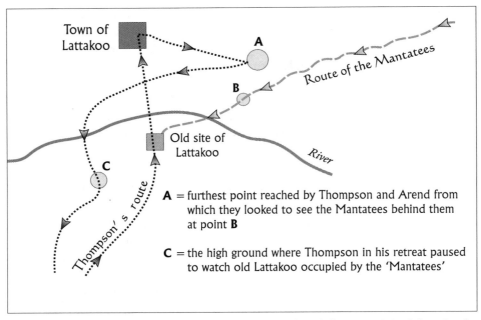

Thompson's encounter with the 'Mantatees' at Lattakoo (after an original drawing by George Thompson)

black mass' in the valley and Arend wisely warned that they should remain absolutely still. This succeeded and the crowd moved past heedless.

Thompson was not yet satisfied. He proposed to 'try the horses' mettle' and 'to gain a front position on the rising ground above the former site of Lattakoo'. This they managed to do quite easily.

There were a few huts left on the old site. At the same moment that the 'savages' rushed into these huts 'like hungry wolves' they looked up and noticed Thompson and his companion, who were not unreasonably apprehensive about their safety. In their retreat Thompson and Arend paused once again to inspect the encroaching mass and moved off just before a group of them, coming unperceived up a ravine, reached the spot.

Thompson refused to speculate on the multitude's numbers but they were clearly many. They were dressed in ox-hides, hanging doubled over the shoulders, though in battle the men were nearly naked. On their heads they wore round cockades of black ostrich feathers and carried war-axes, spears and clubs.

On reflection Thompson realised how the thin edge of time can so easily deliver either life or death. Had he arrived in the area an hour later he would have ridden right into the new town, mistaking the intruders for the Tswana inhabitants. Had he been a few minutes earlier at the point when Arend first spotted the invaders, their march would have been concealed in the windings of the valley.

Thompson and Arend jolted their horses into a pace just short of a canter and rode towards the sun now fast approaching the horizon, beyond the boundless plain of Lattakoo. At Arend's station, they changed the exhausted horses, which had carried them 80 miles that day, for fresher mounts. Thompson, anxious to warn Kuruman, rode the intervening 20 miles in a few hours at a gallop over countryside lit by a bright moon in a cloudless sky.

His tidings were a shock. The Tswana had been tardy in sending out scouts, the invaders were thought to be much further off and they and the missionaries had been lulled into a false sense of security. Since the Griqua commando had not arrived, flight seemed the only option and valuables were hastily buried in gardens. Kuruman was on the verge of being abandoned, perhaps forever.

Saturday was a day of uncertainty – the unknown enemy did not come, but nor did their Griqua friends. The sabbath brought some joy – the commando of about 99 horsemen at last rode in. At an afternoon service when Moffat preached from *Psalm 20: 7*, the church was full. Afterwards it was converted into a fort. Gunpowder was in very short supply and each man was provided with about 12 shots.

Thompson left for the Cape but Melvill joined the commando which set out at 9 a.m. on Tuesday, 24 June, with Moffat in tow in case a chance was presented to parley. The main body halted at the Matlhwaring River while 13 scouts, including Andrews Waterboer (as the Scot Moffat called him, no doubt for the moment conflating him with the patron saint), and Moffat rode four hours beyond.

The scouting party proceeded with all speed at sunrise, and at 10 o'clock they espied the main body of 'the enemy' lying in a declivity between the old and new towns. The scouts stood at a distance, horse-bridles in their hands, trying to convince the confused enemy that they themselves were not intimidated nor did they intend intimidation.

While a messenger was sent to the commando still 20 miles off, Moffat and his companions watched as the vast multitude gathered their cattle and hid them in their midst – this was after all their wealth and their salvation. Occasionally a single warrior would make a rush on Moffat's group but these turned out to be no more than mock charges. Moffat decided to advance to within a 100 metres of the main body and he himself went even closer, unarmed, in order to encourage some dialogue. Their antagonists soon put paid to this plan – they broke out in 'a hideous yell' which stunned the ears and an armed body stormed forth throwing their weapons with great gusto.

The allies retired at some distance, unsaddled, shot a couple of 'wild turkeys' (guinea fowl?) which they buried in the warm ashes of a fire, and hunkered down watchfully. The day petered out without much further excitement.

On Thursday, 26 June, the commando, having ridden up, came in sight of their adversaries at sunrise and advanced to within 150 metres of them. The dread howl arose once more, and the right and left wings of the multitude were flung out in an attempt at encirclement. Missiles were thrown and the 'black dismal appearance and savage fury' of the attackers, calculated to daunt, prompted the commando to take a few steps backwards. Waterboer fired, bringing down a warrior. When a few more were cut down, the wings pulled back. But they were not cowed: a pattern was established – they would charge and the Griqua would fire haphazardly and retreat. But the intruders would never push their pursuit more than a couple of hundred metres from their cattle. Although the Griqua's fire was random it nevertheless exacted a heavy toll amongst the more prominent warriors.

But ammunition was in short supply so the commando changed its tactics. They resorted to tempting small sections out, then galloping in between them and the main body. This proved an effective and lethal deterrence to further incursions. Attempts to capture the cattle were firmly resisted – they seemed to hold their cattle 'dearer than life itself' – and the warriors mingled constantly with groups of women as human shields, which made accurate shooting difficult for the Griqua, and many women were killed.

The Tswana allies pushed forward soon after the beginning of the battle and showered their opponents with poisoned arrows but were easily put to flight by a handful of determined warriors.

After two and a half hours the mass of intruders began to give way. But as they descended into a 'ravine' (probably the banks of the river) they were intercepted with 'considerable slaughter' (although the steep and stony ground made it difficult for the Griqua forces). The retreating crowd then turned northwards and joined another massive body of their compatriots in the new town of Lattakoo. Then the immense united body of mysterious strangers appeared to be taking its departure, moving slowly like a plague of locusts to the north. Cunningly, they tried to use their left and right wings to trap the Griqua amongst the now-burning houses of the town. This was easily repulsed but the scene was dramatic – clouds of smoke rolled overhead and mingled with the dust of the departing crowd. The Griqua harassed the retreat for some eight miles, impressing on the defeated the folly and futility of any further incursions.

Moffat was disgusted by the scenes behind the vanguard. The Tswana, 'like voracious wolves', moved in on the stragglers and 'began despatching the wounded men and butchering the women and children', severing heads from bodies for the sake of a few paltry rings. He was deeply affected by sights of mothers and children 'rolled in blood' and more than once seeing a living infant clutched in the arms of a dead mother. When some women saw that mercy was to be found within the ranks of the commando itself they sat down, bared their breasts and cried: 'I am a woman. We are women.' Sometimes it took 50 Tswana to bring down a defiant wounded warrior. And sometimes the wounded, in the last throes of life, would struggle up and try to extract a final revenge. The battlefield was strewn with discarded karosses, utensils, ornaments, weapons and victuals. These the Tswana rapidly scavenged.

The whole battle had lasted about seven hours. The victory was overwhelming. Moffat estimated the intruders to number between 40 000 and 50 000 including women and children. Between 400 and 500 were slain that day. Only one Griqua was slightly wounded and one Tswana killed while plundering.

On the ride home to Kuruman Moffat was a much less worried man than when he had left a few days before. He pondered the mysterious but wonderful working of providence that had allowed a minute band of 100 horsemen to repel a force of 500 times its size, and the miracle of the battle which would allow him to work at Kuruman for nearly 50 years in the future.

The battlefield today makes for an intriguing visit, perhaps because it is possible to see roughly where the action must have taken place but, at the same time, leaving plenty of scope for speculation. The entire battle, after all, was spread over several miles. I was lucky enough to be guided there by the recent director of the Moffat Mission, Richard Aitken, and his wife, Jane Argall.

The modern town of Dithakong, spread over the sites of old and new Lattakoo, can be approached from the N14 in the south by two roads, one to the west and one to the east of the Moshaweng River. These roads meet at a little bridge over the river at Dithakong, forming a kind of horseshoe. The easiest approach is probably from Lykso on the national road (68 kilometres from Vryburg and 58 kilometres from Kuruman). Dithakong is 30 kilometres north of Lykso on a good dirt road. Just before the bridge on the left is the early twentieth century LMS church and then a little further, again on the left, is a trading store which is probably where old Lattakoo was centred. New Lattakoo was probably across the small river to the west and north. The Griqua coming from the west must have encountered the 'horde' around old Lattakoo or to the west of the river.

The most dramatic part of our visit was yet to come. On a ridge on the west bank of the river, behind a large school which is easily identifiable, is the house of Mr Nkokong. He took us on a short trek of a few hundred metres into the veld beyond. There he pointed out to us a few anonymous stones and a patch of ground identical to anything in the surround. He was adamant (and local tradition supports him) that beneath the innocuous surface lies the mass grave of the victims of the battle. I did not necessarily disbelieve him. If he is correct (and an archaeological dig might clear up a few historical mysteries) then this spot might well be close to the epicentre of the main fighting.

There are two other sites worth visiting in the area. A few kilometres beyond Dithakong lies the village of Bothithong. I had a particular desire to go there. When the first missionaries of the Paris Evangelical Mission Society (PEMS – the French sister mission to the LMS) failed to establish a mission amongst the Bahurutse because of Mzilikazi's intransigence they fell back onto Motito. The fine stone church of Prosper Lemue and Samuel Rolland still stands there to this day, on a rise within gently sloping and undulating surroundings, curiously and sadly not, as far as I know, a national monument. In the cemetery some distance away is the grave of Revd. Jean Frédoux killed in a bizarre incident (of an exploding wagon) caused, according to his tombstone, by 'the misguided resentment of a depraved European'.

An even more dramatic and intriguing site lies just behind the more modern LMS church at Dithakong. There is a huge area of connected stone dwellings (often circular in shape) indicating an ancient ruined town of perhaps several thousand inhabitants. When the LMS first arrived at Lattakoo the locals seems to have no knowledge of these previous inhabitants. In his book *The Colonisation of the Southern Tswana* Kevin Shillington suggests that these were constructed in the sixteenth or seventeenth century. I had never heard of the ruins, had no idea that I was about to see them, and came away with a feeling of some awe. They are definitely worth a visit.

CHAPTER TEN

HOUDENBEK

In the 1790s the tidal wave of revolutionary ideas from France rolled over the shores of the Caribbean islands. The island of Hispaniola was particularly affected. A great slave revolt led by the charismatic Toussaint l'Ouverture in the western part of the island eventually led to the independent state of Haiti. The story is vividly told in that wonderfully inspirational book *The Black Jacobins*, by C.L.R. James.

In 1831 a rebellion broke out near Jerusalem in Virginia. It was on a much smaller scale and achieved very little other than the provision of a martyr to the anti-slavery cause. The story is told in the controversial and somewhat turgid novel *The Confessions of Nat Turner*, by William Styron.

It is often forgotten or only half-acknowledged that South Africa, too, has a slavery past. Does the country have the equivalent of either of these historical events? It does indeed.

The setting for the sad and savage episode lies some two hours' drive to the north-east of Cape Town in wild and wonderful landscape.

You can get to the Ceres Valley through the magnificent Bain's Kloof and Mitchell's passes. The town is, of course, appropriately and beautifully named after the Roman goddess of agriculture, centred as it is in a countryside of fertility and abundance.

The area has been well tilled by writers, too. When Henry Lichtenstein came there at the very beginning of the nineteenth century it was 'at the utmost extremity of the colony to the north', only recently rendered immune to incursions from 'the solitary tracts haunted by Bosjesmans' beyond. The climate being temperate, apples, pears, plums and cherries throve there and the cattle and sheep were particularly fine.

William Burchell's accounts were more personal than Lichtenstein's and got closer to the inhabitants. In the winter of 1811 he took the road towards Karoopoort to what is now the town of Sutherland and was entertained in a remote farmhouse by a slave-owner with incomprehensible arguments, ludicrous dancing and hospitality that put 'more polished society' to shame.

The eminent novelist Anthony Trollope described Ceres nearly 70 years later à la Dr Johnson as 'a real Rasselas valley' but much broader than its literary prefiguration, 'a smiling spot, green and sweet among the mountains', though he suggested that the spirit of prophecy as contained in the name had not yet been fulfilled. The hotel was run by a delightful old German lady and was marred only by the continuous singing of six or seven dozen canaries she was keeping for sale in the front sitting room.

The area has also been well served by contemporary writers. To get a feel for the eighteenth-century context of the hardships of farmers, slaves, Khoi, outlaws and drosters (fugitives) – essential knowledge to understand fully the landscape – Nigel Penn's *Rogues, Rebels and Runaways* is a must.

The other book you should not do without if you visit is *The Forgotten Highway,* beautifully illustrated with the photographs of Paul Alberts. The text by Dene Smuts, the well-known parliamentarian, consists of extracts from many writers (such as E.E. Mossop and Lawrence Green) and fascinating interviews Smuts conducted with numerous locals. The only problem is that this fine book, like so many others in this country, is not readily available – the only freely floating copy in the Johannesburg library system when I tried to get access to it had been taken out over nine months before by a woman from Vanderbijlpark who refused to return it! South Africa is so profligate with its finer publications, allowing them to drop into obscurity.

The valley also boasts one of the last of that dying breed – the small-town hotel owned and run by a Jewish family which used to be so commonplace throughout the rural areas. The Hamlet Hotel, in Prince Alfred Hamlet, is, under the proprietorship of Jos Kahn, in the direct tradition of open generous hospitality that Burchell found two centuries ago and which provides an ideal base for the destination I had in mind.

Which was up the Gydo Pass.

My guide was the former member of parliament and senator Brian Bamford, a knowledgable Virgil to my dismal Dante. Near the top of the pass is a viewpoint with a spectacular prospect. Below is the Ceres valley or Warm Bokkeveld. Above is the Koue Bokkeveld.

While we stopped for hot coffee in the cold air, a couple drove up and engaged us in friendly conversation. The old man was wearing a large tartan beret so Brian asked him was it the real thing. The closest I've ever got to the real thing, came the reply in a cheerful Cape Afrikaner accent, was a tart in trouble.

The Bamfords have a family saying taken from a great heavyweight boxer. When Gerrie Coetzee, once floored by a right hook, was asked what counter he'd devised against it he said, 'I saw it coming and thought, oh, oh, here comes trouble.' On the Gydo Pass Brian had similarly walked straight into the punchline. It was a warning to keep our wits about us in the place where we were going.

On the high ground of the Koue Bokkeveld one of the first things you notice is the rugby fields on the farms, some even with floodlights. Just before the tiny town of Op de Berg is Skurweberg School where Breyton Paulse learnt his rugby. People around here are enormously proud of him. At a nearby garage we asked if he ever came back. Yes, we were told, to visit his mother who works on the farm of the man who has mentored him and subsidised his education. Had he been recently? Yes, he stopped for petrol yesterday in his monogrammed car. Did he give a tip? Yes, R10. Clearly the local hero's success is having, if rugby fans will forgive the phrase, a knock-on effect.

Just beyond the dorp the road branches off to the right taking the traveller to the valley of the Van der Merwes. Very soon one is faced with a flat-topped mountain with an uncanny resemblance to Table Mountain with its own Devil's Peak, Lion's Head and Signal Hill. Below this Tafelberg lies the elegant farm of Boplaas.

A national monument, most of its buildings date from the eighteenth century. It was granted to Izaak van der Merwe as a grazing farm in 1743. Generation after generation of Van der Merwes have lived there ever since. 'My sons who farm there now are the eleventh generation in the country, and are the ninth generation on the farm,' the owner Carl van der Merwe told Dene Smuts in 1988. 'My first grandson and namesake will be the twelfth generation.' A study has found that although the farm Voorhuis near Swellendam has been in the hands of its owning family for the longest time in the entire country, Boplaas has been in the same family for the most number of generations. Boplaas was also part of the dramatic scene of the slave revolt, as we shall see.

There have been other dramas there. One was domestic. In 1894 Carel van der Merwe fell in love with Lizzie Lubbe from far-away Clanwilliam. The trouble was she already had a suitor. So Carel set about winning her with a series of love letters, moving in their simplicity and in their incidental documentation of the changing valley life. Eventually Carel did succeed in his usurpation and he married his beloved Lizzie. But she died young. Being of a sentimental turn she had kept his letters: he did not keep hers. And her letters were found many years afterwards by his descendant Carl van der Merwe and they have been preserved in Dene Smuts' book.

There is little room for sentimentality in the Koue Bokkeveld. It is a stern and unforgiving land. The Skuweberge which run up one side of the valley are hard, serrated – everything has an edge. Only the dassies, modelling themselves on Breyton Paulse, move with a quick agility and elusiveness to evade all black eagles and other predators.

Beyond Boplaas one road leads to Calvinia, past the sand beds and cold waters of Rietvlei and up the desolate Katbakkies Pass to the Karoo. The road to Citrusdal, on the other hand, passes Sneeuwkop and Zuurvlakte. Pilots flying into Cape Town, I am told, take their bearings on Sneeuwkop and start their descent from there. If you have a good guide and a 4 x 4 you can find the track which leads up the Zuurvlakte heights, past leopard spoor and grotesque rocks with shapes like prehistoric monsters, past eerie and abandoned huts of goatherds, past the famous Bushman painting of the yellow elephant to a natural amphitheatre that overlooks the harsh canyon of the Grootrivier.

It is a world of changing moods. While I was there on the edge of winter one sunrise was benign, the next was brutal. There are no in-betweens. It is a country for those who like the wild and uncompromising.

And it produces characters who are equally definite. Probably that is why Lizzie Lubbe chose Carel over her more 'civilised' suitor. And that is what turned one of the Van der Merwes into the unusual poet Boerneef.

'I don't suppose you know where the Dwarsberg and Katbakkies and the Rondeberg and Speserykloof are,' Boerneef wrote in 1948. 'After all, you're a city slicker who seldom wanders away from the highways of civilisation; you're never seized by the desire to flee to places where a motor vehicle is almost never seen, to those areas which are still rough and wild. But you should know about the old Swartrug and Katbakkies and Rietvlei's drifts of sands where people got stuck, till they were left for dead in the olden days …'

These places that he knew were stone-hard, stone-cold, stone-sober and his poetry – so much more concise and allusive than prose – was the same.

> *Met my klipplate klipbanke en pilare van klip*
> *met my kliprante klipkoppe en bakens van klip*
> *is ek kliphard en koud met my preekstoel van klip*
> *my saf kry is min met my hartstreek van klip.*

The present owner of Boplaas now lives in Op de Berg with a magnificent view of his beloved Tafelberg from his verandah while his sons run the farm. Though a hard-headed businessman, it is no surprise that Carl van der Merwe, direct and with a twinkle in his eye, also now writes poetry. He has just published his sixth volume to join a volume of memories.

His poetry is rooted in the local and the colloquial, beyond the full comprehension of an interloper's formal Afrikaans. He is locally known as the 'klein Boerneef'.

At present the inhabitants of the Koue Bokkeveld are complaining. Their main crops are apples and pears, onions and potatoes. There is competition from new strains of apples, costs are high and export costs have to be paid in dollars but are often retrieved in pounds or deutschmarks. But they are survivors over many generations.

Carl van der Merwe knows this. If he is not street-smart he is certainly pad-slim. And the poem he writes about the Hamlet Hotel is a good illustration of what he calls the general *'besigheidkop'*.

> *Die baie gawe Jos Kahn*
> *Hamletse hotelier van faam,*
> *vra my hoe larit deesdae gaan?*
> *(Invra leer lieg)*
> *'Ag man, dit knor;*
> *lyk my*
> *ons ou boere lot*
> *bly nege-en-neëntig persent*
> *bankrot …'*
> *(Ek koop my drank ook daar*
> *met geldjies waarvoor ek geswoeg het,*
> *nadat hy en Manuel*
> *waarde toegevoeg het.*
> *O, ek geniet*
> *die afslag van die 'Israeliet'!)*
> *Ek verluister my*
> *en lag my dood*
> *toe die op-en-wakker Jood*
> *my verder vra:*
> *(dis sy laai)*
> *'Dalk is dit bekend,*
> *maar …*
> *wat van die ander*
> *nege-en-neëntig persent?'*

The book *Bokveldse Binnevet* is on sale at the Hamlet Hotel.

Of course the intensity of the region, coupled with its contradictions, sometimes spills over into violence. Such was the case in 1825.

The times were unsettled. Slaves in the colony were beginning to sense intimations of emancipation which planted expectations in their hearts, and some determined to take their own freedom. An abortive revolt at the farm Wadrif had already provided a precedent. A few kilometres from Boplaas lay Houdenbek, also farmed by a Van der Merwe. He totally trusted his own slaves, even allowing them firearms. The leaders of the new insurrection were the slaves Abel and Galant. On hearing a disturbance in his cattle kraal one night the farmer Nicolaas van der

Merwe opened his door to investigate. He was confronted by his own slaves and was immediately shot dead. A visitor was dispatched in the drawing-room and the farmschool teacher was killed as he tried to hide in the fireplace. The farmer's wife was also hunted down.

A kitchen slave snatched up a two-year-old and fled to Boplaas while another child, aged thirteen, ran for the horses. Unfortunately, he chose a slow horse and was ridden down on what is now the farm Môrester. He was tied to a tree and used as a target by a young slave just learning to shoot.

But when the slave woman reached Boplaas with her news, a small commando was raised from the surrounding farms and it converged on Houdenbek. Galant broke from the reeds along the river on the farm and scaled the nearby Vaalkloof mountain, defiantly mooning his pursuers as he did so. For a couple of days, Galant, Abel and another slave Isa evaded capture in the difficult terrain of the mountains, but on the third day they were trapped in a cave and smoked out.

Abel and Galant were taken to Worcester where they were tried, hanged and beheaded. On the road to Ceres there is a small hill where the heads were placed on stakes on opposite sides of the road as an example to other would-be insurrectionaries. The hill is known locally, in memory of the event, as Koppieshoogte.

Clearly the revolt was very different from that in Haiti where the sugar plantations each had hundreds of slaves and often absentee owners. A revolt on a mass scale like that of Toussaint l'Ouverture was much easier to organise. The Cape was more like the American South where smaller owners seldom owned more than a handful of slaves. One difference from the South, however, was that there was a frontier in the Cape and a hinterland, however unfriendly, for the runaways to escape to and groups of people beyond the border to which the drosters might attach themselves. In the Houdenbek case, however, the events were very localised, and the fugitives for reasons known only to themselves stayed in the vicinity.

I spoke to Naas van der Merwe, the owner of Môrester and Houdenbek. He recalled the week 20 years ago when he hosted a writer who wanted to tell the story. 'He was very quiet,' said the genial Naas, 'and had some strange ways like writers do.' But, as a result, André Brink published his novel *A Chain of Voices* in 1982. Its technique of linked monologues provides the double meaning of the title. The novel fictionalises, but it is based on the actual events and gives a good sense of the countryside.

But to give the novel body, there is no substitute for a visit to Houdenbek in person.

These farms – Boplaas, Môrester, Houdenbek – are immaculately kept and their inhabitants extremely hospitable. You can stay at a very comfortable guesthouse on Houdenbek, which now has a dam teeming with water birds. Until a few years ago the front door of the farmhouse still had the bullet holes in it from the revolt but all the eighteenth-century buildings have gone – except one, which was the ultimate goal of my pilgrimage.

Fortunately, that lone survivor is the slave house where Galant and his fellows spent most of their lives. The first thing you notice is that there were no windows, only two low doors allowing egress. A building without eyes. The dwelling is unassuming but, with the weight of history in your imagination, somehow portentous. I shivered in the sun. From the small building your eyes lift naturally up to the distant mountains. Having scrambled round them I already knew how hostile they are, how inimical they are to the sustenance of life. Only then does one realise how desperate the life of the slaves must have been for some of them to choose the almost certain death in the mountains rather than the cruel life in the valley.

If a relationship to the countryside can be equated with a love affair there are many places in South Africa you can go to for warm affection. But if it is real passion you want then it is to the Koue Bokkeveld that you must take yourself.

CHAPTER ELEVEN

—•—

KHUNWANA

*K*hunwana delenda est! Khunwana must be destroyed! That was the gist of the instructions Mzilikazi, king of the Ndebele, gave to his warriors during the winter of 1832.

He was beside himself with anger at the Tshidi Barolong. He had suspected them of complicity in a Griqua raid on his cattle. Now they had killed the two tax-collectors he had sent to chasten them. So on 6 August he decided to strike at their capital which lay 120 kilometres to the south of his military camp at Mosega.

One wing of the attackers under Gundwana swept round the town to the west; another under Nombate circled round the east side. Their aim was to link up in the south while the third column under Gubuza headed for the royal palace.

In Khunwana, early that morning, two scouts, whose names were Sesedi and Motswapong Lekgatla, raced into the royal kgotla and announced to Chief Tawana that the Ndebele were at hand, having destroyed a Barolong village and captured large numbers of cattle south of the Molopo River the day before. The news was not entirely new. Bahurutse refugees had begun filtering in a couple of days before.

The royal herald was immediately deputed to call the Barolong warriors to assemble at the kgotla, and to warn the people of the danger and the direction from which it might come. This was his cry:

> Hello. Hello everybody. The enemy has been seen.
> The Matabele of Mzilikazi. At the eye of the Molopo.
> All regiments – Old Guards and fresh conscripts – to
> the Kgotla immediately. By the command of the King.

Immediately the men went for their spears and battle-axes, their light shields and knobkerries, and hastened to their stations, while the youths drove the cattle away from where the attack was expected.

They were barely assembled when regiment after regiment of the Ndebele, competing with each other in their show of athleticism, brandishing twin assegais and broad shields, chanting their war-cry of 'Mzilikazi' and emitting strange high hisses, charged the centre of the Barolong 'like demons'.

Despite their disarray, the defenders managed to repel the first assault from the south, the Maabakgomo regiment under Motshegare Tawana and the Malau regiment under Mokgweetsi, son of Phetlhu Makgetla, in particular distinguishing themselves. The latter was later given praise:

> Thou buffalo of thick prominent brow and violent charge
> Thou buffalo that once charged a Tebele
> Made of him mince-meat and mixed him with his excrement
> So that none but hardened fighters could contemplate.

The 17-year-old crown prince Montshiwa, commander of his Mantwa age-group, was also prominent in the fight but the Magalatladi regiment under Sebotso Montshosi was in the thick of it all. Sebotso led the way, killing six of his foes until he took six spears himself from Gobuza's men.

Now the horns of Gobuza and Gundwana linked up at a critical moment and the battle was over, the Barolong fleeing to evade the united Ndebele crescent.

It was more usual for the Ndebele to compel surrender from the youths they conquered and to incorporate them into their own ranks. This time, however, the slaughter was indiscriminate and ferocious – men, women and children. Total war. Butchery. Soon Khunwana was in flames.

Five of Tawana's wives, including Sebudio (mother of Montshiwa) and Mosela (mother of Motshegare and Molema) and two of his sisters, were killed. And amongst the hundreds of Barolong casualties were most of the bravest of their fighting men.

It was Dr Modiri Molema in his description of the battle who adapted Cato's classical adage concerning Carthage to Khunwana. But it was Solomon Plaatje who gave the battle its epical status when he began his fine novel *Mhudi* with an account of the sacking of the town.

I had long wanted to go on a pilgrimage to the major points of importance in the journey that was Mhudi's. In August I was at last able to fulfil this ambition and it was appropriately inevitable that my own journey started at Khunwana, and in winter.

Khunwana is in the Northern Cape, not too far from Mafikeng where Plaatje first made his name during the siege and only a few kilometres from Kraaipan where the first shots of the Anglo-Boer War were fired and an armoured train was captured within sight of the station.

The first noticeable thing about Khunwana is that the surrounding countryside is treeless, almost scrubless even, and pretty flat, offering no natural defences to a swiftly moving Ndebele army. Dust is probably what the Barolong first saw of their approaching foe and dust is what accompanies you as you drive into the town.

Today Khunwana is an unremarkable African town built on a slight ridge with a decidedly unprepossessing kgotla and the sleepiness of a seven day sabbath.

Not the stuff of epic, you might say. But that is the power of the creative imagination. After all, Homer and Virgil made an insignificant dorp on the coast of Asia Minor into the battleground of gods and heroes. So Plaatje breathed life into the dust and ashes of the Kalahari and created a spectacular metamorphosis from the mundane.

Many of the Barolong nobility perished in the battle. A few, including Chief Tawana, escaped the coils of the Ndebele python and made their way south to the temporary protection of their cousins, the Seleka Barolong, under Chief Moroka at Motlhanapitse (near present-day Warrenton).

The fictional hero and heroine of the novel, RaThaga and Mhudi, also flee south, but into the wilderness, where they wander separately for months until they meet and fall in love in one of the great love stories of South African literature. They make their way past Mamuse (in the Schweizer-Reneke district) and cross the Lekwa (Vaal) River, probably close to present-day Bloemhof, so it was to Bloemhof that my pilgrimage next took me.

It is curious that so undistinguished an area has inspired two of South Africa's greatest books. Charles van Onselen's *The Seed is Mine* is the unforgettable story of a black sharecropper who negotiated his way through the intricate maze of climate, politics and custom as one of the great survivors of the twentieth century. Sol Plaatje's *Native Life in South Africa* acutely chronicled the devastating consequences of the 1913 *Natives Land Act*.

In the first week of July 1913, Plaatje arrived by train at Bloemhof station and spent several days bicycling through the countryside, talking to African families who had been forced by the law to take a momentous decision: either to become mere wage labourers, or to leave the farms where they had been sharecroppers or tenants and to take to the roads with their livestock, not knowing where to go or to whom to turn for help.

The cruellest twist was that the Act came into effect in mid-winter. Plaatje tells several stories of the piteous suffering that he personally witnessed as he travelled to Hoopstad. One, particularly poignant, must suffice here.

He spent a night in the open, in a cutting blizzard, with the family of a man called Kgobadi. This man's goats had been to kid when he was forced to trek from the farm he lived on. In halcyon times these kids represented the interest on his capital: now they were dying as fast as they were born and being left on the roadside for jackals and vultures.

Worse. His wife had a sick baby when the eviction took place. Its condition worsened with the jolting of the wagon and soon it died. It had to be buried, but when and where? The stricken parents had no right to any land so they had to bury it beside the road, secretly, in darkness. Plaatje commented bitterly that under the cruel Land Act even little children were sometimes denied the right of burial in their ancestral home.

Plaatje's eyewitness journeys were not irrelevant to *Mhudi*. Quite the contrary. The cruelty of the Land Act is the undertow beneath the historical surface of the novel. Plaatje's journey is also Mhudi's journey, starting from the battle at Khunwana.

There is something deeply disturbing about a concealed burial. As I drove between Bloemhof and Hoopstad the grave of Kgobadi's child haunted me. Was I passing it now? Or now? Or now?

An unknown birthplace is not quite so devastating, though there was an edge to Plaatje's voice when he crossed the Vaal into the Free State and noted that he was close to the district of Boshof where he had been born. 'We remember the name of the farm,' he wrote, 'but not having been in the neighbourhood since infancy, we could not tell its whereabouts.' In fact, he was born near a hill called Podisetlhogo on a farm called Doornfontein.

I decided to visit the farm. The trouble was I didn't know exactly where it was. Plaatje's biographer, Brian Willan, and I had managed to track it down a quarter of a century ago, but memory is a shady customer, offering dubious goods before disappearing into darkness. I remembered that Doornfontein had been subdivided and the name Ghalla Hills had somehow stuck in my hamerkop nest of a brain, alongside the shadowy shape of a telephone tower on a hill.

But when I telephoned all the police stations and local historians in the area none of them had heard of Ghalla Hills. And when I drove south of Christiana on the Boshof road nothing clicked. In the Eastern Cape there is a farm with a white rock which has a habit of disappearing, only to reappear in moments of supreme ominousness. Had I, in a parallel moment of extreme historical carelessness, lost Plaatje's birthplace? Would I be assigned to that circle of hell reserved for sloppy scholars?

Just as I resigned myself to damnation, the telephone tower materialised, 28 kilometres south of Christiana. The farm Ghalla Hills has no sign, no one remembered the historical importance of the place, and no one had heard of Podisetlhogo. Only a small boskampie is still known as Doornfontein.

The farmer's family received me with great hospitality though with some nervousness lest a land claim follow! I assured them that this was less than likely since Plaatje's family were simply transients in 1876 and Plaatje himself had been there no more than months as a newborn baby. Sadly, there is no plaque or monument to mark the birthplace of one of South Africa's greatest. There should be, in case it is lost again – a small token redeemed from the oblivion of forgetfulness.

After their time in the wilderness Mhudi and her family decided to turn east, to go to Thaba 'Nchu. Moving along the Modder River they passed close to where the city of Bloemfontein would be founded within a few short years.

In December 1912, Plaatje was in Bloemfontein for the founding meeting of the African National Congress and he was elected secretary-organiser. (In the tough times that followed he held the organisation together.) In August and September 1913, he was in the city again to see for himself the effects of the pass laws on black women and to talk to some of the 600 who had marched on the municipal offices (scores had been arrested).

But Bloemfontein has an even closer association for *Mhudi* and it was the railway station I had a special mission to visit.

An attractive mixture of Classical and Romanesque, the station was opened to a great fanfare in 1890. The town was lit with lanterns and a great feast was held accompanied by wine, German and English beer, three kinds of whisky, cognac, liqueurs and hours-long speeches from Cecil Rhodes and other dignitaries.

No such fanfare greeted Plaatje's daughter Olive (named for Olive Schreiner), when she arrived there in mid-July 1921. Olive was training to be a nurse in Natal but, her health already weakened when she unselfishly tended to victims of the Spanish Influenza as a young girl in 1918, she had taken ill again and was sent home to Kimberley. While waiting for her connection at Bloemfontein her condition suddenly worsened.

Needing to lie down she was refused entry to the waiting room; then she was forbidden to rest on one of the benches also reserved for whites. She died there on the platform of Bloemfontein station, at the age of 17.

Plaatje's grief at the death of his favourite daughter was deepened by distance: he was in Detroit at the time, trying to draw attention to the plight of his people so far away. The bitter irony was that it now included his own daughter.

Through the museums, libraries and archives of Bloemfontein I tried to locate her burial place but, like that of Kgobadi's child, it eluded me. Olive does have a fitting monument, however. Plaatje dedicated *Mhudi* to her, the victim, as he wrote, of 'a settled system'. The whole book is, therefore, hers.

It is only at Thaba 'Nchu that the importance of the battle of Khunwana becomes fully apparent. After the battle Chief Moroka, the Seleka Barolong and the remnants of the Tshidi decided that Mzilikazi was too close for comfort and they moved eastwards across to the Black mountain in 1833. The move was fortuitous for the migrating Voortrekkers. Three years later, though they managed to repulse an Ndebele attack at Vechtkop (Vegkop), they were stranded there when the Ndebele took off with all their cattle. Moroka and the Barolong came to their rescue with oxen and gave them shelter on the south side of Thaba 'Nchu, at Morokashoek, where the Voortrekkers were to elect their very first government, before moving off to defeat the Ndebele (again with Barolong help) at Mosega. Khunwana, Vechtkop, Mosega – the three battles are inextricably linked like Siamese triplets.

So much greater then was the Barolong sense of betrayal when the Land Act was imposed, and Plaatje witnessed this outrage himself when the Secretary of Native Affairs, Edward Dower, failed to offer any tangible consolation at a meeting there on 12 September 1913. The symbolic effect of setting significant sections of *Mhudi* in Thaba 'Nchu is unmistakable.

Unmistakable, too, from many miles away in the approach from the west is the shape of the mountain, an icon of the South African landscape. In times of drought the people would gather at its foot while the old chief would climb to the top to seek help from the spirits who lived there. Soon rain would fall.

Aside from the Mhudi connection, I had two personal nostalgic reasons for returning to the town. It was there that, nearly 30 years ago, we discovered the gramophone record which Plaatje made in 1923 and on which he sang *Nkosi Sikelel' iAfrika* – the first recording ever made of the national anthem. And it was there that I had interviewed Dr James Moroka, one of the first African medical doctors, president of the ANC in 1949, and a friend of Plaatje's.

He died in 1985 but I now drove to his house just to pay my respects. I was greeted by his son-in-law, Chilly Ramagaga, who offered to show me around the town. When he got into my car, noticed the books and maps scattered about, and smilingly said, 'This isn't your car, it's your office,' I knew I was on to a good thing.

For a whole day we explored some places that I knew, some that I didn't: Archbell's mission house, the oldest European building north of the Orange; the impressive kgotla; the graves of Chief Moroka and his son, Tsipanare, murdered by his brother; Dr Moroka's surgery; and the three Boer monuments (difficult to find without a guide) at Morokashoek.

Above all, the day ended delightfully and unexpectedly as so often happens in such research. Chilly took me to meet an old lady, Mrs Bahumi Motshumi, who had boarded with the Plaatjes while attending school at Kimberley. She remembered that, whenever Plaatje returned from his frequent travels, he gathered the family together to sing the hymn 'A Band of Hard Pressed Men are We'. She sang it for me.

She was also at Kimberley station in 1932 to see Plaatje off on his final journey to Johannesburg where he died, so she was one of the very last to see him alive. He was limping, she told me. Just for a moment, as she talked, I could feel his presence in the room.

Thaba 'Nchu is the furthest east that Mhudi herself travels before eventually turning for home but the story of the making of the novel does not end there. In September 1913 Plaatje went on to Wepener not far from the Lesotho frontier.

Standing on a small koppie early in the morning he stared over the Caledon River into the small and poor country which had generously taken in many of the Land Act refugees and it was with a feeling of relief and freedom that he gazed towards the majestic tops of the Maloti mountains silhouetted against the rising sun.

Did he ever actually visit Lesotho? He did indeed, and in a way significant for the making of his novel. We now know that the bulk of *Mhudi* was written after March 1919, and completed by August 1920. Plaatje addressed the Synod of the French Protestant mission at Morija in October 1918.

The very first novel ever written in Africa by an African was by Thomas Mofolo, who was employed at the Book Depot of the mission. It was called *Moeti oa Bochabela* (Traveller of the East) and was published in 1907.

There is no direct evidence that Mofolo and Plaatje ever met, but Plaatje – who kept abreast of all African publications – must have known the book and his providentially timed trip to Morija would undoubtedly have brought it happily back to mind, a recollection in tranquillity. It may even have stimulated him to start, or at least continue, *Mhudi*. Both books are structured on a journey to the east and Mhudi's journey is a tribute to Mofolo, a resurrection even.

So Morija should be the end of a Mhudi pilgrimage. The old church and the old book depot are still there at the peaceful mission station, silent witnesses to the meeting, spiritual if not literal, of the two great African novelists, and to the birth of the African novel.

Khunwana, Ghalla Hills, Bloemfontein, Thaba 'Nchu, Morija: stations of the cross in the journey that was Mhudi's.

CHAPTER TWELVE

SALEM

Beer and Salem may seem unlikely companions. But the amber ambrosia played its part in many an epic battle on the cricket fields of Sidbury, Sevenfountains, Bathurst and Salem in the 1960s. They invariably took on a predictable form.

On the one side were we callow university students from Grahamstown, naive, gullible and malnourished from residence food (thin gruel flavoured only by the occasional dead frog and prepared by a cook straight out of Chaucer's *Prologue*). On the other side were hardened farmers used to seeing through the wiles of Makanna or the machinations of Lord Charles Somerset.

Invariably we batted first ('Farmer Brown has a gammy leg and can't take to the field except in the early morning') and we would be shot out in no time on a matted pitch which reared and spat like a cobra in must. Then we would be fed.

Course after course served by the farmers' wives with apparently maternal concern, culminating in seven or eight sumptuous desserts including steamed pudding and sherry trifle. And beer.

For the rest of the afternoon we would be mercilessly thrashed around the field, our semi-comatose state disturbed only by the odd torpid puffadder with whom we shared the outfield.

Beer and Salem have a more specific connection, however.

Salem is a small and beautiful village 20-odd kilometres south of Grahamstown. On 18 July 1820 Hezakiah Sephton and his party of 70 families of British settlers arrived at this spot on the Assegai River.

'It is not easy to describe our feelings at the moment, when we arrived,' wrote Revd. William Shaw. 'Our Dutch wagon-driver intimating that we had reached our destination, took out our boxes and placed them on the ground. He bade us 'goeden-dag' or farewell, cracked his long whip, and drove away leaving us to our reflections. My wife sat down on one box and I on another. The beautiful blue sky was above us and the green grass beneath our feet. We looked at each other for a few moments, indulged in some reflections and exchanged a few sentences: but it was no time for sentiment, hence we were engaged in pitching our tent and when that was accomplished we moved into it our trunks, bedding, etc. All the other Settlers were similarly occupied, and in a short time the valley of the Assagaay Bosch River, which was to be the site of our future village, presented a lively and picturesque appearance.' Shaw himself suggested the name of the settlement – 'peace'. In 1822 he built a church of daub and preached while standing on an empty ammunition chest, with his Bible resting on his writing desk which was mounted on a flour barrel. That the existence of the serpent of evil was more than metaphorical he was reminded of on one occasion when a member of his flock cried out during one of his sermons, 'Sir, there is a snake at your feet.' (No doubt, it was its descendants that troubled us in the field so many years later, a sign of the eternal nature of the devil and his remorseless passion to influence the course of a match.)

Ten years later Shaw's rudimentary church was replaced by a more substantial building and at one end of it William Henry Matthews started a school which was to become the renowned Salem Academy (one of his pupils was Theophilus Shepstone who was to play an important role in Natal and Zululand in the years to come).

In 1834, however, peace came to an end for the inhabitants of Salem. The Sixth Frontier War began and the Xhosa invaded Albany. Frightened farmers hastened to the relative safety of Grahamstown and Bathurst. Some made for Salem where Jeremiah Goldswain, in his inimitable spelling, recorded seeing 'English Dutch Hottentots Bushuners etc.' barricade themselves inside the church. Others erected temporary defences around the church from wagons, boxes and stones. Peace had left them unprepared.

The first Xhosa to arrive, early in the new year, took off with a large number of cattle under cover of darkness. Next afternoon on the hill known as Monkey Kop facing Salem from the north appeared a small army of Xhosa some 500 strong, clearly intent on attack. The situation for the defenders, poorly armed and with no immediate hope of rescue, was desperate. The oppressive heat was ominous.

Beer was a fishing village, chalk-cliffed and the ancient haunt of smugglers, on the coast of Devon where Richard Gush was born into the Episcopal Church in

1789. He went to London at the age of 21 and converted to Methodism under the influence of the Wesleyan family he lodged with. Married to Margaret Evans in 1811, he sailed with the Sephton party on the *Brilliant* in 1820. He soon found he had to supplement his income from carpentry by trading. One of his treks took him into the Karoo and he was disturbed at the treatment the colonists meted out to the Khoi, the Bushmen and to their slaves. He also heard a mysterious voice say to him three times, 'Return to Salem and preach the Gospel.' Initially he ignored these promptings but when he had his stash of money stolen and discovered, on his return home, his house and goods washed away in a flood, he soon divined the cause and the effect. In addition, he came under the influence of the Society of Friends.

In the scattered raids running up to the full-scale war that saw his cattle stolen and his house, built on a rise behind the church, raided, he refused to retaliate. When danger threatened he refrained from joining the beleaguered villagers but stayed in his house, going about his business. But when he spotted the Xhosa on Monkey Kop that day in early 1835 he was seized with a divine righteousness and, despite his wife's anguished pleas, decided to ride out to meet them to try to prevent the seemingly inevitable bloodshed. He took off his coat to indicate to the Xhosa he was unarmed. With him went his daughter's husband, Philip Amm, and a burgher called Barend Woest.

Through the valley and up the hill they rode, no doubt to the puzzlement and astonishment of their adversaries. At an appropriate distance Gush dismounted, held out his arms to indicate once again that they were empty and invited two of the Xhosa to approach him in peace.

Forthrightly, with the help of one of the warrior's interpretation, he chided the commander, indicating that the missionaries William Shaw, Stephen Kay, Simon Young, John Ayliff and Samuel Palmer had all come to them via the Salem Mission. 'In our church,' he told him, 'we pray for you that God may show us how to help you and that He may guide you to live better lives. Ever since the white man came to Salem, missionaries have visited you, doing good. Why do you threaten us?'

'My people are hungry,' was the reply.

'How is it possible,' Gush retorted, 'when you have taken all our cattle?'

'Yes, but we want bread,' said the commander.

Gush promised them bread in return for an undertaking to depart in peace. He rode back to Salem, picked up two loaves of bread weighing seven and a half pounds each, ten pounds of tobacco locally grown at Salem, and a dozen pocket knives. Not enough to feed the entire band, no doubt, but enough to satisfy their leaders. Saluting Gush, they turned away. It had been an extraordinarily courageous act by Gush, a combination of faith and heroism.

It would be interesting to know the equivalent motives for the Xhosa restraint but, of course, the sources are lopsided and one can only guess at their magnanimity on this occasion.

Gush died at Woodbury near Salem in 1858. His descendants live in the house he built to this day. A room or two and verandahs have been added but its original outlines can still be seen. In one room a couch made by Richard's son, Joseph, can be found and on the gates to the house the words 'Devonshire' and '1832' announce its ancestry. In front of the old church is a plaque commemorating Gush's deed and if you stand in front of it and look straight out over the fine cricket oval you can see Monkey Kop on which there is a similar plaque on the spot where the incident took place.

The ghosts of many battles no doubt haunt the oval, and the clash of competing sides are heard on misty nights, but inside the beautiful third and newest church, it is worth sitting for a while and taking in the memorial window to two much greater wars.

In the silence you can hear the ticking of a clock made by the Grahamstown clockmaker H.C. Galpin. It is a timepiece of thanksgiving.

In 1868 little George, the three-year-old son of Samuel Shaw, the headmaster of the Academy, went missing. A search at night revealed nothing. The other dangers, snakes and wild animals, were an ever-present threat. The next day and the next night, followed by yet another day and night, passed in fruitless search of the surrounding countryside. On the third night a thunderstorm dampened all hopes for the child's survival.

On the fourth day a cattle-herd and tracker in the employ of Mr Shaw, 'his keen eyes accustomed to following game trails', spotted a small footprint in the veld, then a tiny shoe. The searchers crossed the opening, formerly known as 'Fish Lane', now 'The Glade', near what was once called Vasco da Gama Road. The child was found, dehydrated and starving, but he eventually recovered on a diet of milk in small spoonfuls. The tracker was duly rewarded with a cow and calf and the clock has been quietly ticking away for over 130 years.

On a day when the sun is at its highest and the heat its fiercest the church, with its white walls and high ceiling, is a perfect haven, a place worthy of its name.

CHAPTER THIRTEEN

CONGELLA

On a wall in the town hall at Grahamstown there is a plaque commemorating the end of the epic ride which Dick King made from Port Natal in 1842. On his horse 'Somerset' he had ridden over 1 000 kilometres in ten days, bringing news that a British force had been defeated by the Voortrekkers at the Bay and was threatened with capitulation.

If you travel through the Eastern Cape you can pick up indications of his journey along the way. For instance, outside Peddie on the main N2 road to Grahamstown, there is a turn-off to the north to Committee's Drift. Two kilometres along the road is the beautifully restored and kept fort at Trompeter's Drift. It is now a private farm though the hospitable owners wryly joked to me that, because of poachers and their dogs, they wished 'the soldiers were still there!' Just before you reach the farm and fort there is, on the left-hand side, a stone marker pointing out that the intrepid horseman had passed that way.

Similarly, on the war memorial that casts a cold eye on the bustling and chaotic main street of Butterworth is an equivalent sign. It was a ride which undoubtedly had a profound effect on South African history.

Permanent settlement by whites on the shores of Natal had begun in 1824 when Lieutenant Francis Farewell, Henry Francis Fynn, James King and others established themselves precariously as wary traders and land concessionaries, nervously keeping a watch on the Zulu power which surrounded them.

The whole picture changed when the Voortrekkers moved east of the Drakensberg in 1837 and eventually set up the independent Republic of Natalia with its capital at Pietermaritzburg. The British, who regarded the south-eastern seaboard as within their sphere of influence, watched suspiciously. In fact, in December 1838, they sent a military force under Major Samuel Charters to occupy Port Natal. Charters chose a position on Point St Michaels (one of the two arms, together with the Bluff, which, like a nearly closed pincer, formed the entrance to the great bay) to build a stronghold which he called Fort Victoria after the new sovereign. But a year later the British withdrew leaving the Voortrekkers in possession of Natal.

But the Voortrekkers, with lemming-like zeal in provoking the British, pursued a policy of marking out large farms and stirring up the local tribes, not least of all by raiding them for cattle. Hugely controversial was their attack (with the encouragement of Faku, chief of the Pondo) on the village of Ncaphayi, chief of the Bhaca Xhosa. Since this was south of the Umtamvuna River and therefore in an area he himself claimed as his domain, it made Faku, not worried apparently by contradictions, decidedly nervous and he asked the British for protection for his territory between the Umzimvubu and Umzimkulu rivers. The governor, Sir George Napier, in January 1842, chose to send a force under Captain Thomas Charlton Smith to the mouth of the Umngazi River (just south of present-day Port St Johns) to oversee the situation. Napier was concerned that pressure from the north would destabilise conditions on the eastern frontier just recovering from the Sixth Frontier War (1834 to 1835), squeezing the Xhosa and Pondo into an unhappy claustrophobia.

In fact, Napier had already decided to re-occupy Port Natal and Smith's force was the means to do it. He would wait for the summer rains and floods to subside and estimated it would take Smith 17 days to march from the Umngazi to the Bay.

His anxiety increased when news came that the brig *Brazilia* had dropped anchor in the harbour at Port Natal. The previous year the presence of the American brig *Levant* had signalled that foreign powers might interfere in regional affairs on the side of the Voortrekkers. Now the *Brazilia* was a more direct threat – the boat represented private interests in Holland eager to take the opportunity to trade with the trekkers and a pamphlet unwisely distributed by the captain raised false hopes in the trekkers that they would be supported by the motherland (the government of the Netherlands hastily disavowed any part in this initiative).

Smith left the Umngazi on 1 April 1842. The bulk of his force consisted of infantry of the 27th (Inniskilling) Regiment, backed by a howitzer and two light field pieces of the 4th Battalion Royal Artillery, a few Royal Miners and Sappers, and a dozen-and-a-half horsemen of the Cape Mounted Rifles (CMR). In all, a force of just over 260 soldiers.

It was encumbered not only by a large and unwieldy wagon train (60 of the drivers were armed Englishmen) but also by a score of wives and children.

Smith, a veteran of the Peninsula War and Waterloo, but by all reports only a moderately competent officer, was presented with an unenviable task. He was advancing into unknown terrain with a force that would be heavily outnumbered and far less mobile than the skilled horsemen and crack shots which were the trekkers. Smith, too, did not understand his opponents and probably underestimated them.

The column set off with the artillery in the lead together with the contingent of the CMR and an advance guard of infantry, followed by the main body of the 27th and with the supply wagons bringing up the rear – the whole force spread out over two miles.

A lively description of the march was provided by Joseph Brown, a bugler whose job was to keep the sections of the column in touch with one another.

They swung out of the Umngazi and marched through the territory of Faku to the confident strains of 'We fight to conquer'. In the following days a Mrs Gilgen (probably the wife of an infantryman) gave birth to a son and shortly thereafter the wife of the commissariat issuer was delivered of 'a beautiful daughter'.

The countryside they trudged through was soaked and their boots sank in deep and every extraction sapped their energy. Brown recorded that they crossed 122 streams and rivers, some of them 600 metres wide. When they came down from the hills to the coast (near modern Port Edward) they gratefully gambolled in the sea and gorged themselves on a cornucopia of shellfish. New sights greeted their eyes: seacows and the skeletons of whales. But the going, in the soft sand of the beaches, was no easier.

The whole march, what with the mud and the sand, the rivers and the childbirths, took Smith much longer than Napier had allotted him.

In Port Natal the English settlers wanted to inform the approaching column of what was going on, and what it might expect. Samuel Beningfield, defying the curfew imposed by the Voortrekker government, slipped away but failed to make contact with Smith. On his return to his home, his excuse about going hunting was not believed, and Andries Pretorius, commander of the trekker commando, ordered him to be tied to a tree and, in front of his wife and children, shot. The marksman aimed, fired and missed. Twice! Pretorius looked down his nose at the incompetent executioner and said that 'if that was the best he could do' he might as well cut the prisoner loose. With that he rode off to busy himself with more important business.

Almost certainly, it was a rehearsed farce. Pretorius wanted to send a warning to the settlers and a bit of theatre was not a bad way of doing it.

On 29 April the British column sighted the *Brazilia* which had not long before left the Bay. The next day Private Devitt died of fatigue. In spite of letters of protest from the Volksraad, on 4 May, Smith's men bivouacked on a hill near Dunn's 'Sea View' estate overlooking the Bay about 10 kilometres away. 'The evening we came here,' wrote Brown, 'we saw the haughty Dutch banner was displayed on the fort at the harbour, as large as life.'

The following morning Smith, with the CMR and some artillerymen, rode to the port and brought down 'the rebellious flag'. The military engineer of the British force, Lieutenant Gibb, deemed Fort Victoria to be 'totally untenable' and chose a slightly elevated piece of ground 800 metres away for a camp (more or less on the site where the Old Fort now stands and right next to Kingsmead cricket ground). The entrenched position consisted of a ditch and earthen walls, with the wagons arranged in laager formation on the outside. There was a fairly good sweep of fire for the 24-pounder howitzer, the 18-pounder field piece and the other smaller guns. The water supply was adequate and secure though James Brown complained: 'Such a place for bad water I never saw before in my life, it is as black as ink and full of different insects and stinks into the bargain. I am afraid it will make away with the whole of us before long.' It was not an ideal position: it was at a distance from the harbour and vulnerable in its supply lines from the sea, and it was out of eyesight of the entrance to the Bay.

The irrepressible bugler was soon fretting at his own confinement: 'No person of our camp is permitted to go to town since we came here, we are locked up the same as if we were in a French prison.'

The trekkers under Pretorius ('six feet high with a belly like a bass drum', Brown said of him) were camped at the village of Congella on the shores of the Bay to the south of the British camp. Their numbers increased steadily until they outnumbered their enemy by well over two to one.

In the following days negotiations, with Smith demanding submission and Pretorius insisting on British withdrawal, got absolutely nowhere. Smith's supplies were critically low, though the arrival of the small brig *Pilot* (on 13 May) and the 90-ton schooner *Mazeppa* (on 20 May) brought some relief.

On 23 May Pretorius raised the stakes. As the latest deputations were conferring (Revd. Archbell was interpreter) Pretorius was arranging for his men to take possession of the grazing British cattle as soon as the talks broke down. Up to 600 head were seized. What he intended is uncertain but he was well aware that Smith needed them if he was to retreat overland with the wagons or as a supply of food. Perhaps he was hoping that Smith would withdraw in the two small ships still riding at anchor offshore.

Smith decided he must act. He could wait and defend an entrenched position. In hindsight this was probably the better option. Instead, he chose to be preemptive. He prepared to attack Pretorius's position at Congella. And at night.

His main force, under Captain Lonsdale, consisted of about 100 men of the 27th. Two ox-drawn six-pounder field pieces, under Lieutenant George Wyatt, with an escort, were in the vanguard.

There were two possible routes for the advance. The first was diagonally across the area which is now the CBD of Durban but it was broken up with clumps of bush friendly to the hidden trekker vedettes. The second route was southwards to

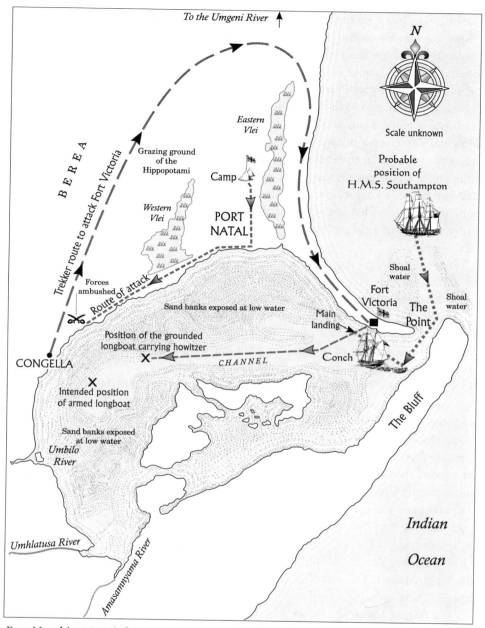

Port Natal in 1842 (after a contemporary sketch by Lieutenant Gibb, RE)

the northern beach of the Bay followed by a swing westwards towards Congella. This route was longer but more level and open than the first.

Smith chose the latter option. He wanted 'to cross the sands at low water, as by doing this I could avoid annoyance from the farmers'. He was to strengthen his artillery support by mounting the 24-pounder howitzer on one of *Mazeppa's* longboats, which was to link up with him as he approached Congella.

It was a mild autumnal Durban night with a light south-westerly breeze blowing. There was a full moon in a clear sky. At 11 p.m. (Monday, 23 May) the main body set out in three divisions, each of two ranks of sixteen soldiers in each rank and with the guns leading. Smith rode in the company of two CMR troopers and two buglers. Their feet sank ankle-deep in the soft sand and mud and sometimes they waded knee-deep in the water, but they held reasonably good order. One of the field pieces made a noise from a frictional hasp in the stillness of the night.

Surprise therefore did not last long. The trekker vedettes were soon alerted and by half past twelve Pretorius in Congella knew of the development.

The longboat carrying the howitzer was also in trouble. The sergeant in command found the boat falling behind his comrades marching along the shore so he ordered an increased rate of stroke from the rowers. This led to fewer soundings being taken so that the boat ran slap bang into an unseen sand bank that jutted out into the Bay from the stream which ran along what is now Field Street. The chaos that this caused meant that the boat never did participate in the action. Whether the howitzer on such an unstable platform would have been in any way effective and fulfil its objective of driving back the trekkers is debatable.

As the British moved awkwardly through a mangrove swamp, the trekkers, in concealed position, opened up a deadly fire from their elephant guns (which outranged the soldiers' muskets). 'I was forced to make the attack,' Smith subsequently wrote in a dispatch to Napier, 'without the valuable assistance which a discharge of shells and shot from the howitzer would have afforded me.' The six-pounders seem to have got off a couple of volleys but wounded oxen running amok soon upset the limbers and delayed reloading.

The trekker fire was deadly. Lieutenant Wyatt, RA, was hit in the forehead and died immediately. The Inniskillings returned fire but, caught in the moonlight, began to take casualties steadily. The situation was hopeless.

The retreat, led by Smith, was effected under difficult conditions, the trekkers pressing closely upon the fugitives. Much to their shame the British had to abandon their field guns. Smith, who got back to the camp first, organised its defence as best he could. Sure enough, a large body of the attackers converged on the laager and 'opened a heavy fire on three sides of it'. This was kept up until an hour before daybreak, when Pretorius finally ordered a pull back. It is possible that he acted more cautiously because of an attack in the rear by a small party of settlers under George Cato.

There was a lull in the fighting that day (Tuesday, 24 May). The British had lost

17 dead (4 artillerymen, 11 soldiers of the 27th Regiment and 2 CMR troopers) and some 31 with wounds of varying severity. Sadly, the liveliest voice of the whole expedition was silenced with the death of Private James Brown, bugler of the 27th. The trekker casualties were much lighter: perhaps one, perhaps a few killed and a handful wounded.

Smith decided to remain in his camp, although many of the women and children (20 in all) were moved to the Point and boarded the *Mazeppa*. Smith could now not get away either by land or sea. His only chance was to be relieved by an outside force, and his only hope was to get a message away.

At midnight George Cato woke Dick King on the *Mazeppa*. Cato and his brother Joseph, Peter Hogg and John Douglas took a boat and towed two horses across the Bay, King and his servant Ndongeni holding the reins as the horses swam. They landed at Salisbury Island and the party was seen and fired on, Cato being captured. But King and Ndongeni slipped away, armed with a couple of pistols each, waded through the mangroves, crossed the Bluff and rode along the coast to Umgababa, hiding during the day and forcing themselves onwards at night. Thinking that only King was taking the message only one saddle had been provided by Cato and so the 16-year-old Ndongeni had to ride bareback.

On Thursday a party of about 100 trekkers left Congella, rode under the frown of the Berea and swung right round the British camp to its west and north and advanced on the Point. The fortifications there were defended by Sergeant Barry and 18 men, 2 gunners and an 18-pounder. They were supported by a few settlers. The sentry at the flagstaff was shot and the attackers arranged themselves in the cover of the sand hills and poured fire into the barracks.

A servant of the trader Ogle swam out to the *Pilot* but had got just halfway when he was shot in the back of the head. The small garrison, realising their position was futile, gave themselves up to the trekkers who assured them they would not be harmed. The captured civilians were taken to Congella, then to Pietermaritzburg where they were put in gaol, some in the stocks. The captured 18-pounder was then turned on the ships, which also surrendered, though oddly their crews were allowed to stay on board.

In the days that ensued the trekkers spent their time plundering the stores at the Point and pushing their trenches around the camp while those besieged tried to strengthen their defences. On the 31 May the besiegers attacked the camp with the captured cannons and during the day fired 122 shots into it. Thereafter they settled in for a long siege.

Conditions in the camp were serious. Deep trenches had to be dug to protect the men against the cannonballs. Medical supplies were almost nonexistent and the wounded had to be accommodated in hollowed-out recesses. Horses had to be killed for food and supplies rationed. Water turned brackish and dysentery and enteric followed.

A couple of night-time sorties were made by the British to discomfit the denizens of the trenches, which crept ever closer on three sides, and some hand-to-hand fighting took place with bayonets (British) and gun butts (trekkers). There were some casualties.

The besiegers' grip tightened, the food was increasingly rationed, the wounded and sick suffered. On 10 June a small event gave some hope. Joseph Cato persuaded the crew of *Mazeppa* to attempt to escape. They barricaded the side of the ship with mattresses, and with the help of the women and children, hoisted their sails. Then they made a run for the entrance of the Bay, pursued by hot fire from the elephant guns on shore. The mattresses proved effective and the schooner escaped. It headed for Delagoa Bay where it was thought it might run into a Royal Navy warship.

By 25 June 651 shots had landed in the camp. The defenders were down to a small lump of horseflesh per day, a handful of biscuit crumbs, and forage-corn ground into meal. Eight soldiers had been killed in the siege. The tide of resistance was running low, made worse by the fact that Smith did not know if King had succeeded in his mission.

On the night of 24 June, almost exactly a month after King had left, a rocket appeared in the east. The besieged answered with their own. During the next night almost a fireworks display of rockets rose up above the sea.

The first boat to arrive was the *Conch*, packed with the grenadier company of the 27th Regiment from Port Elizabeth under Captain Durnford and skippered by Captain Bell, who knew Port Natal well. They were concealed below decks, in hatchways and anywhere else that offered a modicum of cover. The Port Captain, who approached the ship, was taken completely by surprise at the nature of the ship's cargo. He was persuaded to take a letter to Pretorius asking for permission to land a doctor to help Surgeon Frazer at the camp: Pretorius denied the request, forbidding any communication with Smith.

Next evening, the frigate *Southampton* appeared offshore. She had hastened from Cape Town on hearing the news of Smith's distress and carried a large detachment of the 25th Regiment under Lieutenant-Colonel A.J. Cloete.

The following day, Sunday 26 June, would be critical for the future of the whole region. Would the British establish a permanent presence in Natal?

Pretorius spread his men along the strategic points of the coast. Commandant Louw Erasmus took up position with 100 men at the mouth of the Umgeni River. Ocean Beach was patrolled by about 40 horsemen. Pretorius stationed himself with a similar number of men at the Point, backed by a four-pounder gun. On the Bluff the Winburg Commando comprised a formidable force of 350 men. Once the direction of the British attack was known Erasmus was to hasten to the aid of Pretorius if necessary.

At three o'clock in the afternoon the *Conch* crossed the bar and entered the channel. She carried 135 troops and towed four boats with 120 men of the 25th. The *Southampton* was 'warped up to the bar' (that is, it was hauled up to it by a rope fixed to a stationary object ashore) and proceeded to fire uncomfortable broadsides at the Point and the Bluff.

Erasmus was too late to back up Pretorius, who resisted for as long as it took him to realise that his line of retreat from the Point might be cut off. So he first withdrew to the trekker lines around Smith's camp, then to Congella. The landing force – which had lost two men killed and six wounded – came ashore at the Point, took over the fort and replaced the republican flag with the naval ensign. They then marched to the camp and lifted the siege.

For good or ill the British were back, this time for good.

The trekkers had not managed to shake the British monkey from their backs. Pretorius retreated to Cowie's Hill after the battle but in July the 25th Regiment, under Captain D'Urban, marched on Pietermaritzburg and on the 15th the Volksraad gave up. Conflict would resume nearly 40 years later.

The ride to rival that of Paul Revere (during the American Revolution in 1775) had taken Dick King ten days (seven days shorter than could have been reasonably expected). After leaving the Lovu River which he and Ndongeni had crossed the first night and then concealed themselves in thick bush, they swam across the Umkhomazi the following night where they were warned by friendly locals that they were being pursued. So they redoubled their pace and by dawn found themselves at Umzimkulu, present-day Port Shepstone).

By now they were in friendlier terrain and felt safe enough to ride by day. They reached the Umngazi but Ndongeni had to give up. The chafing he had endured without a saddle became unbearable. King himself was delayed for two days by illness but pressed on to bring the bad news to Grahamstown.

Ndongeni eventually walked back to Durban, but he was not forgotten. He was given some land on the north bank of the Umzimkulu River, was still alive in 1911, and was buried there at a ripe old age. I do not know whether the grave is marked. (King died, aged 57, on his farm near Isipingo in 1871.)

One hundred years afterwards, in 1942, the ride was commemorated by the journalist and poet, H.I.E. Dhlomo.

NDONGENI
(Dick King Centenary Reflections)

Ndongeni! Praise! Seed of our fatherland!
Son of ancestors bold, you braved the sea
And dangers great to serve and teach this land
Example of the power of unity.
King's deeds and glory are your glory and
Your deeds! His fame and praise, your praise and fame!
The full diapason of that theme – Dick's band,
Can never be complete without thy name.
The beauty and the treasures of your home,
Ndongeni – hear and help! – your progeny
No longer call their own! Your children roam

Despised! Outcasts! And ever do they flee!
Upon our visions doubts close like a mask!
'Was it for this Ndongeni rode?' we ask.

A new nationalism, which would soon lead to the creation of the African National Congress Youth League of which Dhlomo was a founding and prominent member in Natal, was looking for new heroes.

CHAPTER FOURTEEN

— • —

BOOMPLAATS

In August 1848 the astute commander of the emigrant Boers laid a trap for the supremely confident Governor-General of the Cape, Sir Harry Smith. It remained to be seen whether he would fall into it.

At the beginning of the year Smith, who had recently succeeded Sir Harry Pottinger as governor, saw it as his task to bring under British control the numerous squabbling groups (the Boers, the Griqua, the Barolong, the Batlokwa, the Basotho) north of the Orange. The Griqua cottoned on fairly enthusiastically to the advantages of such protection and Moshoeshoe and the Basotho somewhat reluctantly so. The Boers, having had to abandon their dream of a republic in Natal, were more likely to balk at such a move but Smith believed that his previous popularity amongst them and the force of his personality would settle matters decisively. So he undertook a quick-fire trip through the Transgariep area and Natal.

In January he visited both Bloemfontein and Winburg and received a polite welcome so that he was able to present a case to his superiors in London that the Boer population supported him. On the 27th he met Moshoeshoe, his sons and his adviser, Eugène Casalis of the Paris Evangelical Missionary Society. Moshoeshoe was conciliatory but did not want to lose any of his land. Smith did meet Andries Pretorius in Natal and it was agreed that the latter should cross the mountains and feel out the mood of the Boers, though there might have been some misunderstanding over whether this meant only those north of the Vaal River.

On 3 February the theatrical governor, on the banks of the Thukela River in Natal, issued a proclamation declaring the Queen's sovereignty over the whole area between the Orange and the Vaal rivers eastwards to the Kathlamba (Drakensberg) mountains.

He waved away the concerns of Major Warden, the British Resident in Bloemfontein, regarding security, asserting that the Boers 'are my children', and then he hurried away, leaving only a small detachment of the Cape Mounted Rifles (CMR) and Warden in charge of a vast and turbulent country.

Within months Pretorius, tormented by the final illness of his wife, had rallied the majority of Boers behind him and on 17 July he advanced on Bloemfontein with a commando of 1 000 men (200 from the Transvaal and 800 from the Orange Free State). Warden had only 57 troopers of the CMR and some 40 armed civilians at his disposal. He had little option but to agree to terms which would allow him to withdraw to the Cape with personal belongings in tow. On 20 July Pretorius entered Bloemfontein, then moved on to Middelvlei on the north bank of the Orange.

Before the fall of Bloemfontein, however, Warden had managed on 13 July to get off a dispatch to Smith. This report arrived in Cape Town on the 22nd. The Governor wasted no time in responding and ordered all available units to Colesberg. Leaving the Mother City on 29 July he travelled the approximately 820 kilometres to Colesberg in less than 12 days, quite a feat when the fastest mode of transport was the horse.

The Orange was in flood. There was a pont higher up the river (located there some five years before by a Scotsman called Norval) but Smith had had the foresight to bring with him two india-rubber floats (it was 1848!). So Smith could cross the great river at Botha's Drift, some 25 kilometres north of Colesberg. This was completed in five days and the whole force was across by the afternoon of 26 August. Pretorius was not there. Shortage of grazing had prompted him to bivouac on a farm called Telpoort, some 100 kilometres north of Philippolis. He was also motivated by a rumour (unfounded) that a British column was approaching from Natal and he might have to protect his rear should this prove true.

Smith left 40 men of the 91st Regiment and 20 men of the CMR at the ford and marched the 28 kilometres to Philippolis on the 27th. His force consisted of two companies of the 1st Battalion, Rifle Brigade (under Major Beckwith); two companies of the Reserve Battalion, 45th Foot (Captain Blenkinsopp); two companies, Reserve Battalion, 91st Foot (Lieutenant Pennington); four companies of the CMR (Captain Armstrong); a small detachment of royal engineers; and a few 'loyal' Boers. There were three six-pounder guns and a large baggage train under Henry Green. Estimates of the number of men differ but it may have been about 800. In Philippolis they were joined by 250 mounted Griqua who, according to the historian Theal, 'varied in appearance from the pure savage in a sheepskin kaross to the half-breed in plumed hat and European costume'.

On the 28th Smith drove his army north along the Bloemfontein road to Visser's Hoek. If the reader takes himself or herself off to this area he or she might come to the same realisation as I did of how hardy the infantry must have been. They marched 40 kilometres in a day with gun and pack through harsh countryside, flat and rugged, beautiful to the home-grown eye but alien and harsh to the gaze of a foreigner. Then the weary infantryman might have to stand guard half the night with every prospect of having to fight next day.

Smith was familiar with the road to Bloemfontein and the terrain. He expected Pretorius to defend the strongest position along it – a group of hills 25 kilometres north of Visser's Hoek centred on the farm Boomplaats (which was 80 kilometres from Bloemfontein). In fact, Pretorius did precisely that. When his scouts confirmed Smith's exact position on the night of the 28th, he moved his commando of some 800 men south from Telpoort (about 40 kilometres), and set himself up to ambush his barney.

The usually gung-ho governor this time did not want to fight. He had sold the annexation of the territory to his superiors with the argument that the majority of the Boers favoured it. A nasty engagement with casualties would take a lot of explaining. His men, on the other hand, eager to show their mettle, were more ready for a barney.

At dawn on the morning of the 29th the British column pushed forward. It was cool but there was not a cloud in the sky and the air was crisp and bracing. They breakfasted at Touwfontein and then resumed their march over the open plain towards the beckoning hills. In the front rode the Cape Corps (the CMR) with its 'European officers and Hottentot soldiers', in their uniforms of dark green, and their carbines slung at their sides. Behind them, at 'infantry pace' marched the rifle brigade, then the sappers, miners and artillerymen with their field guns. They were followed by the 45th and the 91st. The wagon train trailed behind; the farmers and the Griqua guarded the flanks and the rear.

It is quite easy to get to the site of the battle. Just off the main road (N1) between Cape Town and Johannesburg, 100 kilometres south of Bloemfontein, is the small town of Trompsberg (now home to the writer Karel Schoeman). If you take the R717 to Philippolis just outside Trompsberg you will encounter a dirt road (the R604) leading off to the right towards Jagersfontein. Along this road you will pass the farms Vlakfontein (on the left) and Plaatjiefontein (on the right) and 22 kilometres from the turn-off you will come to a sign pointing to Swartkoppies and Touwfontein to the left (the south). Opposite, on the right an unmarked road takes you 4.8 kilometres to the farm at Boomplaats. As you travel along this road you will be roughly following Smith's route – the old, old main road to Bloemfontein!

The first features encountered are two hillocks (one with a radio mast on top) just to the right of the road. A bit further on is a range of low hills and the road crosses this range some 300 metres from its eastern end (the range is about one-and-a-half kilometres long). The range is not very high and from a distance appears almost insignificant. Beyond it the road dips down to a stream called the

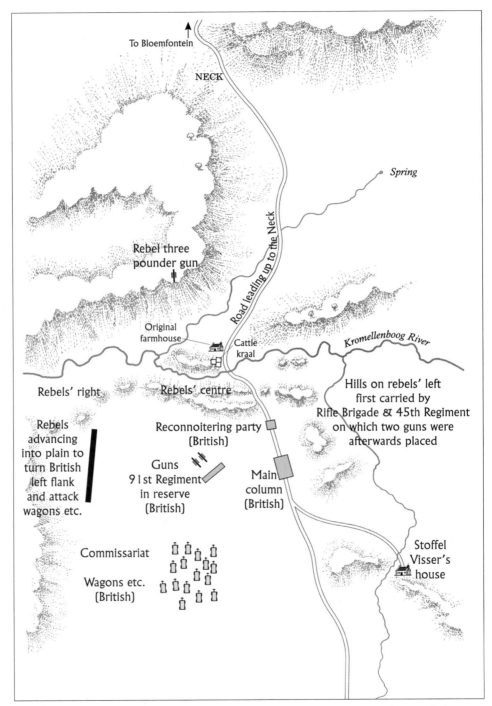

The Battle of Boomplaats (based on a map in Sir Harry Smith's autobiography)

Kromellenboog (or Crooked Elbow). Today there is a low bridge across the chain of pools: in 1848 this was the drift which was crossable.

Beyond the bridge is a gate with a commemorative plaque of the battle. Behind the gate is a distinctive oval koppie, a significant landmark. Nowadays the road runs to the east of the koppie: then, it seems to have bent round its west and north faces. A mile further down the road a range of high hills dominates the landscape, visible from a long way off to anyone moving up from the south. The road curves round east of these high hills to a Neck, the gateway to the north.

Pretorius aimed to use this terrain to set his trap. He placed his sole three-pounder gun on the high hills which dominated the whole position towards the north. His hope was that Smith would assume that this was where his stand would be made. In fact, he concealed his main force behind the seemingly insignificant range of hills south of the stream. He placed his advance guard on the two hillocks that Smith would pass first. It was hoped that at least part of the British column would pass the hidden Boers before they revealed themselves and the fire from both sides would be murderous. If you examine the plan from afar off to the south it makes sense. The high hills in the distance make a more obvious place to defend than the insignificant ones in the foreground and the depression in which the stream runs is not visible.

On the morning of the 29th, then, the Boers took up their main positions and the Transvaalers were in the centre of the low range just to the west of the road, with Free Staters on both sides, Stander being on the end east of the road. Kock was on the west flank, behind a hill, waiting to hit Smith's left flank from the side and rear. A few Boers were positioned on the two hillocks further south on the road. It was as good a plan as one could devise in the circumstances.

But that morning the British chanced upon a lone shepherd who confirmed that the Boers were at Boomplaats. So Lieutenant Warren and a small detachment was sent ahead to scout as the main column approached the hills. At 11 o'clock some of Stander's men, taking up their stations on the advanced hillocks, disturbed a herd of grysbok who streamed across the front of the Boer position. Warren, alerted, rode to the nearest hill and spotted 40 or so Boers there. The trap had been sprung too soon.

Smith, in blue jacket, white cord trousers and drab felt hat, pushed forward a troop of the CMR under Lieutenant Salis abreast of the second hillock, only 50 metres off the road. The Boers in white hats and dusty jackets appeared on the crest and poured a couple of volleys into the hapless troopers. Three of them were struck down and Salis, his horse shot dead under him, had his arm shattered. As he sat in the road, he heard a Boer urge his companion to shoot him. 'You must not,' cried Salis, 'for I have a wife and children.' The Boer shouted to him, 'Are you wounded?' To which Salis replied, 'Yes.' He was allowed to crawl back, and was stretchered off to a field hospital.

A rifle ball had grazed the face of Smith's horse, and another cut one of his stirrup leathers. Cursing profusely as the main Boer position opened fire, he deployed his troops with practised precision. The wagons, guarded by the Griqua, were retired while the CMR fanned out to the left. The royal artillery, escorted by the 91st, opened up on the Boer centre and left while the 45th and the rifle brigade were ordered to charge, the latter in an attempt to outflank the Boer left wing. When they made their way to the top Captain Murray was hit three times and several of his men were cut down.

As the British charged, Kock on the Boers' extreme right rode forward to attack the wagons but was driven back, failed to reconnect with Pretorius and had to retreat and circle right round the high hills in the north in order to rejoin the main force at the Neck, effectively out of action for the remainder of the battle.

The CMR then joined the line with the 91st on their left. The advance of the regulars was so swift that the Boers were sent reeling from the ridge of hills. But Pretorius managed to rally them at the Kromellenboog, using the oval koppie as a central point of resistance, with a stone cattle kraal near the farmstead providing a deadly strongpoint. The small lone Boer gun on the distant heights opened up along the road but to no effect and it was eventually knocked out.

By 12.45 p.m. the Boers were in retreat, falling back towards the Neck where they made a last stand against the CMR and Griqua who, being mounted, could chase them more swiftly than the infantry. For a while the Griqua were held at bay but when the artillery and infantry came up the Boers had had enough and by 2 p.m. they broke and made their individual ways home. They were pursued by the British for several miles until 4.30 p.m. Then the British, who had been marching and fighting since dawn, halted.

Smith gave the Boer casualties as 49, 12 of whom were killed by a single artillery shell. It is difficult to know how he got this number. The Boers always denied that there were ever more than nine dead. Some of their names are on the monument on the top of the oval koppie – Nicolaas du Buisson, Petrus Erasmus, Diederik Lafnie, Douw Steyn, two De Beers (without first names) and two unknowns. Their bodies are believed to be buried at separate and scattered sites around the farm. The oval koppie provides the best, 360 degree, prospect of the whole battlefield, this time primarily from the Boer angle.

Below the very modern farmhouse is a small cemetery containing the graves of the two British officers killed – Captain Murray of the rifle brigade and Ensign M. Babbington Steele of the CMR. Six other men of the rifle brigade (Barret, Day, Daunahy, Thomas, Martin and Hollister), five of the CMR (Mentor, Camphor, Freeman, Jack and Williams), three of the 45th (Halton, Harvey and Baylis) and six unknown Griquas also lost their lives. Forty-six British and Griqua were wounded.

Although the British losses were heavier their victory was decisive. On the day after the battle they were at Telpoort and on 2 September at 9 o'clock in the morning they marched into Bloemfontein.

But five years later they withdrew from the Orange River Sovereignty.

CHAPTER FIFTEEN

⬧

FORT BEAUFORT

T he Queen's Highway from Grahamstown approaches Fort Beaufort from the south. Close to the town it reaches a T-junction (the R63). The road to the left travels past the Kroomie mountains to the small town of Adelaide with its unusual circular town centre. The road to the right leads almost immediately to another T-junction (the R67). Another right turn takes the traveller across the modern bridge over the Kat River at the point which used to be known as Stanton's Drift and into the town of Fort Beaufort.

In this area the Kat River has formed a large horseshoe loop before it swings back on itself to join the Brak River. The town is contained within this horseshoe, its two main streets (D'Urban and Campbell) running parallel to one another.

A left-hand turn at the R67 junction heads north along the valley of the Kat River to Lower Blinkwater, Tidbury's Toll, the Katberg, Whittlesea and, ultimately, Queenstown.

Into this beautiful valley had been allowed to settle, after Chief Maqoma had been pushed out in 1829, a mixed group of people. Some were Khoi; some were Gona (of Khoi-Xhosa ancestry); others called themselves Bastaards (the result of white-slave or white-Khoi parentage).

During the 1830s and 1840s they had founded a thriving settlement with healthy orchards and tilled lands freshly watered. In addition, they had loyally fought for the Cape government in the wars of 1835 and 1846 and had suffered substantial losses from Xhosa raids. By 1850, however, the frontier had moved eastwards and, with the waning of their strategic importance and the need for their military assistance, envious glances were cast on their fertile lands.

They were treated, too, with arrogant contempt by their resident magistrate, T.J. Biddulph. When one of their leaders, Hermanus Matroos, formerly interpreter to Harry Smith the governor himself, complained of the huge quitrent he was expected to pay, the governor threatened him with eviction. Then in June 1850 Biddulphs' successor, Thomas Bowker, sent in a punitive police expedition which burnt down many homes of a people who had proved their loyalty time and again. The leader of the Gona, Andries Botha, wrote to the Governor, Harry Smith, complaining of this treatment: 'Your Excellency knows me; I am an old servant of the Government and I hope a faithful one; I served under Government in the war of 1835; your Excellency knows I never flinched from duty; I never feared to face the enemy – and that with the very men who have now so shamefully been expelled from the Settlement.' Within a few months Smith was to pay for his high-handedness in ignoring these entreaties.

By the end of 1850 the governor and the Cape Colony were in deep trouble. In 1848 the boundary of the Colony had been extended eastwards to the Keiskamma River and the loyal Mfengu were located in a series of settlements in the area. In the Tyumie River valley, the traditional lands of the Ngqika at the foot of the Amatole mountains, four military villages were founded at Juanasburg, Woburn, Auckland and Ely. The area between the Keiskamma and the Kei rivers was proclaimed a dependency of the Crown, named British Kaffraria, and at a great meeting in King William's Town just two days before Christmas Smith humiliated Xhosa chiefs such as Ngika's son Sandile, Anta, Mhala, Siyolo and Phalo by compelling them to kiss his boot while he sat imperiously on his horse and forced them to address him as '*inkosi enkhulu*' (great chief).

This humiliation was exacerbated by the undermining of chiefly authority with its replacement by colonial officials. The loss of land, authority and independence was fertile ground for the mystical prophecies of a young man called Mlanjeni who had acquired his powers from bathing in an enchanted pool. His reputation was enhanced when a great earthquake seemed to shake the foundations of the world and many saw him as a reincarnation of Makanna. Preaching sacrifice, Mlanjeni's words were warlike.

The Ngqika were unhappy. When Smith summoned the chiefs to a meeting on 26 October 1850, Sandile did not appear. Smith was furious and immediately officially deposed him as chief. The Ngqika went almost immediately into rebellion. Ngika's other son Maqoma joined Sandile.

Smith was not in a strong position. In response to the Colonial Office's eternal plea to cut costs he had bent his own knee at the expense of the military and had sent the 27th, 62nd and 90th Regiments and the Dragoon Guards home. The military campaign he undertook was consequently a disaster.

On 24 December the main column sent into the Amatole mountains was severely mauled at Booma Pass by that brilliant tactician Maqoma. On Christmas Day 1850 the military villages of Woburn and Auckland were overwhelmed and Juanasburg was hastily abandoned. Smith, caught in Fort Cox, only narrowly escaped to King William's Town where he found himself besieged and in a precarious situation. Fort Beaufort was cut off from Fort Hare and Fort Cox. The only positive thing, for the British, was that the Gqunukhwebe and most of the Ndlambe remained passive.

Smith was in desperate need of allies. One group he turned to were the Kat River settlers. But arrogance breeds fawners, not friends. And the Kat River people, under Hermanus, made a pact with Sandile to attack Fort Beaufort.

Because I wanted to view the battle at least partly from their perspective, I took the road north from Fort Beaufort along the Kat River and fortuitously found very comfortable accommodation at Baddaford Farm. In the sitting room two paintings of 10th Cutting by a local farmer immediately transported me back into a 150-year-old landscape. My friendly and helpful hosts Jane and Jonathan Roberts put me onto a local historian, Gert van der Westhuizen, who knows the terrain of the whole area so well. Before I met him, however, I wanted to associate the preliminaries to the main battle with some of the sites in the countryside.

While the Ngqika might have taken the government by surprise, they did not have the same effect on many of the farmers, who had heard strong rumours of trouble for over a year. However, these farmers were not prepared for the difficulties that did overtake them. Some 13 kilometres west of Fort Beaufort a wealthy landowner Benjamin Booth lived with his family on the beautiful farm Hammonds, a sturdy structure of stone surrounded by a stone wall and sited in a fine position. With the help of their domestics and a man from the Kat River settlement whom they had brought up since a child, and with a plentiful supply of ammunition, the Booth family were determined to defend themselves and their possessions.

However, when he heard of the military setbacks and the Christmas massacres at Woburn and Auckland Booth took his family into town, just in time. Booth had left his loyal servants to defend the farm. Not for the first time and not for the last did a master discover the ambiguities of servant 'loyalty'. The domestics plundered what they could, destroyed everything else and made their way to join Hermanus and the Kat River settlers.

The Gilberts lived in Sipton Manor. After the destruction of over 400 farmhouses in the war of 1835, the government gave advice on the construction of fortified farms for the protection of people and livestock. George Gilbert applied for permission to build his outside Fort Beaufort not far from Hammonds. It was a well-designed construction, defendable by the owners and their trained domestics, and with an enclosure capable of holding in safety his span of oxen and 500 sheep. There was a fresh spring, fruit and vegetable gardens, and a year's supply of provisions. William Gilbert was determined to defend it but eventually prudence prevailed and he and his family managed to reach Fort Beaufort safely.

Eight kilometres along the Adelaide road (the R63) is the turn-off to Sipton Manor. Two kilometres of good dirt road takes you to the farm itself. Raymond and Jan Pearson, who live there, graciously showed me round this magnificent example of a fortified farm. The great walled stockade with shooting loops all along it is still in place, as is the original horse stable with its manger. The loft is also there where 40 pockets of potatoes and 10 sacks of onions were kept in anticipation of the siege. The house, too, is still there with firing slits overlooking the door. The Pearsons told me of the smoke on the rafters on the roof where flames had done their best to destroy the house and showed me the outlines of the old main road which had once passed just in front of the house. Raymond produced an old lead ball from a *voorlaaier*.

I asked him if he had lived in this area all his life.

'No,' he replied, 'I grew up in the Adelaide area.'

Everything is relative (Adelaide is 25 kilometres away).

Back outside he showed me a couple of water tanks he'd made from metal chests used for transporting tea and keeping it dry. I'd never seen one of those before.

Gert van der Westhuizen lives with his wife Mattie on the farm Olive Cliff near Baddaford Farm on the road to Lower Blinkwater and Queenstown. He took me part-way up Fuller's Pass and pointed out the patch of dense bush high up in a cleft in the mountain which is called Maqoma's Den and from which the chief taunted the British with guerrilla warfare through 1851. At the foot of the pass, beyond the farmhouse, he showed me the camp where Hermanus Matroos had his village and where he pondered his attack on Fort Beaufort.

Tall, sturdy, with dark wrinkled skin and curly hair, Hermanus was about 50 years old. Clearly he was no fool.

On Christmas Day he visited town, trying to buy ammunition. When one storekeeper asked him where he stood in the conflict he pointed to the sky and replied enigmatically that 'God above will tell me what I should do'. He got so far as to approach Lieutenant-Colonel William Sutton (Cape Mounted Rifles [CMR], son-in-law of General Sir Henry Somerset and senior officer of the military forces in the town) for arms, but Revd. John Ayliff did not trust him and persuaded Sutton to reduce his supply of guns to a minimum. Hermanus no also doubt took the opportunity of assessing the town's defences and its armed forces.

Many of the buildings – especially the military ones – stand to this day, superb examples of their kind. At the head of the town (at the opposite end of D'Urban Street from Stanton's Drift) was the large double-storeyed military barracks. There was a blockhouse in front on Somerset Street and a strong guardhouse at one end of the barracks. Further down Somerset Street, on the corner of Bell, was a sturdy Martello tower (which is still intact, only the wooden floors being slowly stolen). On it was mounted a small cannon. Lower down, on the corner of Church and Campbell streets (the building is no longer there), St John's Anglican Church was turned into a fortified point when news filtered in of the setbacks to the east. Beyond the barracks was the Mfengu village.

Hermanus must have noted that the forces at Sutton's disposal were flimsy. There was one company of the 91st Regiment, about 50 to 60 strong, some sappers and miners, and a detachment of the CMR whose allegiance came to be regarded as decidedly iffy. A force of 600 Mfengu was commanded by two white officers. In town the Fort Beaufort volunteers comprised 60 men, and the Fort Beaufort Mounted Volunteers 40. They used St John's as their headquarters. Hermanus believed he could take the town, particularly backed by the element of surprise.

He and his allies had a strategy in place. The plan was for the Xhosa under Sandile to attack the Mfengu village from the north while the Kat River contingent would attack across the three drifts (Stanton's, the one near the military hospital and Johnson's) but not, curiously, over the Victoria Bridge.

Hermanus' line of early march from Olive Cliff on 7 January followed roughly the R67 to Fort Beaufort. A couple of kilometres north of Stanton's Drift he paused to address his force of about 500 fighting men. They were divided into three groups which were to sweep round and approach the drifts assigned to them but wait for the signal to attack which would be the first shots from Stanton's Drift. Brandishing a knife he said it would be used to cut throats and all Englishmen were to be killed. Mfengu men were to be slain and their women taken, while he would have white women serve him tea on the verandah of Holliday's store. Afterwards he would have his way with them. Those of his men who fought hardest, he said, would get the greatest share of the booty and he reminded them of his reputation, his power and his relationship with Mlanjeni, the seer.

At about 4.30 in the morning Hermanus' men crossed Stanton's Drift and occupied several of the houses in the lower end of town. Some even had time to shout to the Mfengu maids in the trader Stanton's house to prepare them a good supply of coffee. But surprise was only half achieved.

The women and children of the town had been gathered in the barracks and had slept soundly that night. Twenty regular soldiers manned the guardhouses and the Martello Tower and a regular watch was kept. The Mfengu were guarding the river banks.

When the first musket shots were heard, Sarah Ralph woke those around her and they went to the windows. On the first floor of the barracks where they'd slept and from which they had a commanding view of the town though it was still dark, they could at first see only the flashes of the muskets.

One of the citizens of the town, William Wynne, was wakened by three shots – he mistook them at first for the guards celebrating the end of their watch but when several more rang out he was startled from his dozing reverie. Grabbing his firearms he rushed out into the street where he met Captain Savory and both of them, with no time to collect their horses, ran down D'Urban Street towards the sound of firing. Others were doing the same. Meeting some Mfengu herding

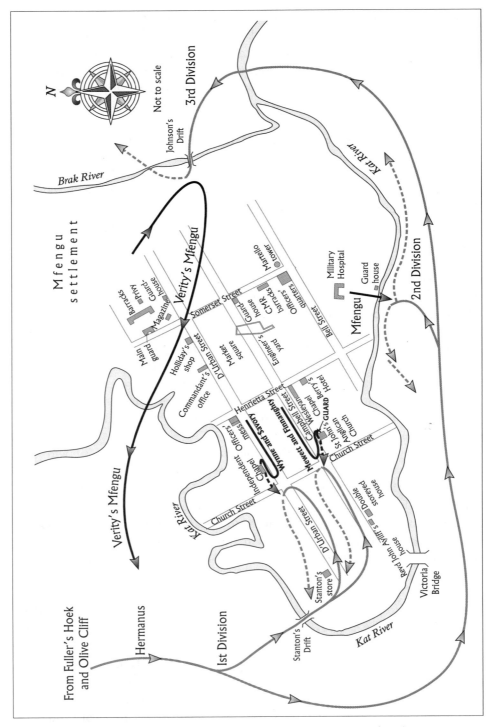

Attack on Fort Beaufort, 7 January 1851 (after a sketch by N. Mapham)

their cattle up the road they shouted at them to join the fighting, which many of them did.

By the time they took temporary shelter near the Independent Chapel on D'Urban Street, Wynne, Savory and others began to return fire with some success. When their position became too hot for comfort they joined some 40 other Volunteers in and around St John's.

The second group of Hermanus' army swung round the town and started their attack across the drift near the guardhouse and the military hospital five minutes after they heard the opening shots.

Sarah Ralph had a bird's-eye view over the hospital from her eyrie in the barracks and as the sky was lightening she could see the Mfengu stationed there attacking the besiegers with such ferocity that they were sent scattering in all directions.

The third group under Hermanus himself swung in an even greater arc further round the town – almost three-quarters of the way round in order to cross the Brak River (almost no obstacle at all) at Johnson's Drift. But by the time they got there those of the second group put to rout were already pressing on their flanks in panic. They were struck head on by a largish force of Mfengu accompanied by several Volunteers under Captain Verity.

In this action Hermanus was killed almost instantly. This took the fight out of the men around him and they fled.

Verity did not pursue them. Instead, he turned back through the upper part of the town, crossed over the Kat River behind Holliday's store onto the flat ground to the west and threatened the left flank of the first group of Hermanus' force.

That group itself was already in retreat. The Volunteers and the Mfengu, having settled themselves into something resembling an organised force along Church and Campbell streets, had poured withering volleys of fire into those of Hermanus' men who had tried to rush Revd. Ayliff's house on Campbell Street near Victoria Bridge.

At 5 a.m., 20 to 30 minutes after the firing had begun, the main battle was over. The Gona and Khoi were in panicky retreat on all fronts. The Mfengu and the Mounted Volunteers – pursuing from Stanton's Drift – and Verity's force threatened to enclose the first group in a pincer movement, while the rest fled northwards so fast they could not be caught by the avenging horsemen at full gallop.

At the hill on Olive Cliff the main body did turn and make a stand but the excited Mfengu did not falter in the face of furious fire and turned them yet again. When the Mounted Volunteers sent a message to Colonel Sutton urgently requesting supplies of ammunition, he sent six troopers of the CMR with 2 000 rounds. Otherwise he and the Regulars took no part in the battle.

At the Blinkwater River Hermanus' remnant force divided. Some of the cavalry stayed at an abandoned post there where they set about taunting their pursuers but were soon dislodged by the Mounted Volunteers. The other section turned towards Hermanus' camp at Fuller's Hoek. But the Mfengu were soon upon them, killing many, even including one in Hermanus' own hut. The Mfengu found a number of wagons stuffed with goods taken from the farms of Booth, the Gilberts and others. These they looted as well as over a thousand cattle and hundreds of sheep.

On their joyous and victorious return to Fort Beaufort they were met, 5 kilometres out, first by Colonel Sutton and the CMR and then, soon after, by Captain Carey, who had been sent by General Somerset from Fort Hare with a detachment of CMR as a relief force. The Mfengu triumphantly entered the town chanting 'Mlanjeni gedoen, Mlanjeni gedoen, Mlanjeni gedoen' (Mlanjeni is finished).

Some 80 of Hermanus' men were killed and many wounded or captured. They were not faceless, anonymous statistics. Some were known. Amongst them were people who had worked for the Aytons and Gilberts: Figland Fransman, an old man beaten nearly to death by the Mfengu; Klaas Botha, lead singer of the Kat River Settlement and son of the veld kornet, Andries Botha; the deacon of Revd. van Rooyen's church and a former employee of Sarah Ralph's father. Of the latter Sarah said something most interesting, a lesson unlearnt – he was, she said, 'one of the last anyone would have thought would have become a rebel'.

Captured Ngqika prisoners revealed that Sandile and an army were in the area but Hermanus had made a grave error and had attacked too early so their joining up was never effected, otherwise the battle outcome might have been more in doubt.

In the event the battle was decisive. Only two Mfengu were killed the whole day and all whites emerged entirely unscathed. Having proved their mettle the Mfengu were allowed to keep their booty to the chagrin of some of the farmers, who had to endure, in the days to come, seeing many of their best clothes finely displayed on darker bodies.

Sutton suggested a thanksgiving service to Ayliff, but Revd. Wiltshire refused to allow St John's, the largest and most suitable church, to be used for such a purpose. Despite remonstrations from his own congregation, he was adamant. 'It is quite against the rubric,' he said. So the service was held at 5.30 p.m. in the Wesleyan Chapel but was joined by many Anglicans.

During the morning after the battle round the town was over, while the dead and wounded were being identified, the body of a well-built man was found lying some 200 metres east of Johnson's Drift. It was Hermanus.

The body was stripped and thrown over the back of a horse. It was taken to Market Square where, under the market bell from which a Union Jack was flying, Sarah Ralph saw it placed, naked, wearing, bizarrely, only a woman's crêpe bonnet.

The ignominious display of the body of Hermanus was, announced the local newspaper, 'a warning to TRAITORS'.

Which was an indication of how little they understood him. Or, indeed, what had happened.

CHAPTER SIXTEEN

CENTANE HILL

The road to Qolora Mouth from Kentani is not paved with good intentions. When Kaizer Matanzima was in power and Transkei was for a few years an independent state the president took money from the local people and promised them a tarred road. It never happened. Or perhaps it did.

The locals will not admit to being duped and, blessed with a fine sense of humour, they will tell you that it is indeed paved but they have covered it over with dirt and grit and sharp stones to protect the fine surface beneath. This is comforting. It is also nice to know that our leaders would never consider stealing from the people.

Which is one way of saying that the road to Qolora Mouth (and there is a pont across to Kei Mouth) is easily passable in an ordinary car but it is not a smooth road. A puncture is a likely happening. There are some hazards along the way, too. Not just the usual cattle and ox-carts (there are some fences but these seem to be designed to keep the cows and oxen out of the field and on the roads) but also a healthy number of black pigs. They are there for a reason. Because many of the households use middens rather than gold-plated flush lavatories with doilies or fun-fur round the rims, the pigs clean up. Before they are slaughtered, however, they are penned in for two weeks and fed mealies and greens to flush out their insides.

At the end of the road there are two welcoming hotels – Trenneries and Seagulls. The latter has a sign in its friendly pub which tells you that 'vegetarian' is an ancient Native American word meaning 'lousy hunter'.

The first objective of my journey to Qolora Mouth was to see the mysterious and mystical pool where a 16-year-old maiden saw her visions.

Some time in April or May 1856, Nongqawuse, niece of a seer-prophet called Mhlakaza, was sent by her uncle to chase birds away from a cornfield. Her village was situated high above the Gxara River mouth. She and another younger girl, Nombanda, decided to slip down the hillside and bathe in a deliciously cool pool below. Suddenly two mysterious men appeared before them, giving names of warriors long dead, and instructing them to convince their people that, if all their cattle were killed and all their cornpits emptied and all witchcraft renounced, 'new people' would arise and fill new kraals and new pits with cattle and corn and the whites and the Mfengu and any unbelievers would be pushed into the sea.

Next day she took her uncle and others to the pool and although they could not see the strangers who reappeared to the maiden they could hear what they said through her.

Word spread and eventually Sarili (Kreli), paramount chief of the Gcaleka, and indeed of all the Xhosa, came to stand on the high hill and was shown the phantom of his own dead son and herds of horses or cows in the sea.

The Xhosa had heard that the British were at war with the Russians in the Crimea and expected the Russians to come and deliver them. So, many did kill their cattle and emptied their cornpits and in 1857 they died in their thousands of starvation. But the pool does have a magical and historical pull and it has for many years been my intention to make a pilgrimage to it.

To get there you definitely need a 4 x 4. You also need a guide. You need Trevor Wigley of Trevor's Trails and Adventures. Behind the hotels and small village a desperate track takes you to a hill and a stunning view overlooking the sea and the mouth of the Gxara River. The village on this hill is called Baku and this was the village in 1857 of Mhlakaza, Nongqawuse's perhaps visionary, perhaps scheming uncle. Trevor can point out Nongqawuse's *tsholo* (little wood) where the orphan girl lived. From here she would descend to the pool.

The trouble is there are two rivals for the honour. The one closest to the sea is more authoritatively regarded as IT. It has a grassy bank, a pool and a krantz on the other side. The second one (on the other side of the hill and further up the river) is more closed in by indigenous bush and a terrific krantz and has an air of mystery and meaning about it which surely would attract the more powerful spirits. This is intensified by the presence, hidden within the twining embrace of dense bush, of a Khoikhoi tomb or monument dedicated to an ancient god.

It is certainly an experience to stand on Baku Hill and know that this was where Sarili stood and, through swirling mist, saw horses or cows in the sea and the spirit of his own departed son.

There are a number of explanations for the phenomenon of Nongqawuse. One is that her experience was the result of a psychotic state of mind of a young woman. A second, in the light of the fact that several other prophetesses were active at the time (the most notable being Nonkosi who went further than Nongqawuse by immersing herself in a marshy pool on the Mpongo River, living with the water spirits and actually meeting the prophet Mlanjeni who had been behind the 1850 to 1851 war), was that she was part of a wider social movement, the result of a society under heavy external pressure. Others might argue that it was the result of woman's ambition or anger or some deeper stirring of the family or social pot. Another version finds its answer in the cynical machinations and manipulations of Mhlakaza (after all he was no naive or unlettered primitive, having lived for some time as 'Wilhelm Goliat' under the powerful, individualistic tutelage and influence of the Revd. Merriman in Grahamstown and had aspired to his own religious vocation). Some have argued that the whole movement was a last-ditch plot by the great chiefs to force the Xhosa into armed resistance as the only alternative once they had deprived themselves of their cattle and grain. Yet others see behind it all the sinister hand of the British officials and the governor, Sir George Grey, hoping to break the back of Xhosa cohesion.

Each version might come from an appeal to whatever interests were or are involved. Probably no single cause is sufficient – a mixture of some or all of the above, as well as others, is likely to be more satisfying.

But, whatever the causes, the consequence of the movement precipitated the greatest social upheaval in Xhosa society. The bare statistics alone tell a story: within a year (1857) the Gqunukhwebe population was reduced from about 8 000 to 650; the Ndlambe from 23 000 to 6 500; the Ngqika from 43 000 to 5 500. Some had sought refuge amongst the Tembu, some had gone to the mission stations and to the Colony for food and for work, many had succumbed to miserable starvation.

But that was not the end of the story, certainly not of the more immediate consequences. Many of the chiefs – Maqoma, Mhala, Xhoxho – were imprisoned on Robben Island. Sandile humbled himself to the governor and was permitted to lead a miserable existence under the eye of Charles Brownlee at Dohne. Sarili slipped into exile across the Mbashe River after fleeing a punitive raid by Walter Currie and the Frontier Armed and Mounted Police (FAMP) into Gcalekaland (David Hook, who was with the invaders, saw a land inhabited only by barely moving skeletons and the Gxara River area, where it had all begun, littered with human skulls). Nongqawuse herself (after a period on the Island) ended her days in Lower Albany and was buried, in 1898, on the farm Glenshaw in the Alexandria district where her simple grave, in a clump of trees, can be visited to this day.

The wretched existence of the Xhosa dragged on for 20 years but Sandile and Sarili were not prepared to give up without a fight. It is sadly symbolic that the Gcaleka (aided by the Ngqika) came to the area of Nongqawuse and Mhlakaza for their last showdown at the battle of Centane Hill.

It is a site seldom visited and it is completely unmarked. So much so that when Trevor Wigley and I went in search of it, we took the wrong track and got lost (not a disaster since the magnificent views of the various valleys and the distant heights of the Kei River were a happy diversion from the fruitlessness of the search). It was therefore with an added feeling of exhilaration that we eventually discovered the exact location of the battle.

It is located on a hill (called nowadays Moldenhauer's Hill) about a kilometre along a track off the main tarred road between the towns of Kentani and Butterworth. Precise and detailed directions are desirable. If you are coming from Kentani there is a green sign on the left-hand side but pointing to the right marked Ndabungele (on the right up a slope is a store with a slogan 'I will be bright' painted on it and there is a lone telephone booth standing on the right as well). The turn-off to the battle site is actually to the left along a rough track. It is only accessible with a 4 x 4. If you are walking or driving along the track you will soon pass a blue and white building of The Gospel Church of Power and will immediately see the deserted shell of Moldenhauer's trading store. Your path should lead along the right side of the store's buildings. Then, just to the right of that is the slope of Centane Hill. It is easily found. On the crown of the small hill are the remains of a fairly large earthwork fort. In front of its walls are long rows of depressions – clearly old trenches. To the south and west the fort occupies a fairly commanding position. To the north there is a wooded valley which might offer attackers some cover though there is still a vulnerable open stretch to be traversed before the fort is reached. To the east, though, the land slopes upwards and the top of this offers a view overlooking the position.

The ninth and last frontier war had begun in August 1877 with a minor incident involving a bloody fight that erupted at a Mfengu wedding which some Gcaleka, as was legitimate by custom, had attended uninvited. The new governor, Henry Bartle Frere, tried to visit Sarili across the Kei but the latter, fearing the same treacherous capture and death which had befallen his father Hintsa, had fought shy of a meeting. Although he wanted to avoid a war the radicals amongst his subjects pressured him into one.

A force of the Frontier Armed and Mounted Police under Charles Griffith was assigned to support the Mfengu and they were encamped at Ibeka (today a suburb of the sprawling modern town of Butterworth). There in late September a patrol was attacked by the Gcaleka and narrowly escaped annihilation. Then Ibeka itself, defended by 180 police and 2 000 Mfengu, resisted an attack, on 29 September, by the Gcaleka army under the renowned warrior Khiva and a young prophetess who urged the attackers to advance not in open formation but in massed, closed ranks. Shells, rockets and Snider bullets sent them reeling: the prophetess was killed and her head cut off by the Mfengu. Another attack the following morning was also repelled.

But panic spread to the inhabitants of the Ciskei and King William's Town was abuzz with rumour.

Frere and the Commander-in-Chief of the Imperial Troops, Sir Arthur Cunynghame, calmly set about securing the situation. The frontier was sealed off by regular soldiers while the colonial militia was to cross the Kei and deal with Sarili once and for all (Frere issued a proclamation deposing him as paramount chief).

On 9 October Griffith surprised the Great Place of Sarili at Holela and burnt it to the ground. On 17 October he swept through Gcalekaland but found it empty – Sarili had once more retreated beyond the Mbashe. Preparations were made to settle the vacant land. Complacency set in; the colonial militia went home.

But on 2 December a fierce fight took place at Holland's Shop when a police patrol, accompanied by some artillery, was attacked by 1 000 Gcaleka warriors under Khiva. The Gcaleka were back and posed a real threat! Another panic followed.

A force of rough railway gangers and diamond diggers was cobbled together by Colonel Henry Burmeister Pulleine, of the 24th Regiment. They were mockingly dubbed 'Pulleine's Lambs'. And Lieutenant Fred Carrington, also of the 24th, raised a force of 200 mounted riflemen. On 21 December General Cunynghame reached Komga ready to join his troops in order to cross the Kei. But before he did so a mixed patrol of cavalry and police under Major Hans Moore of the 88th Regiment encountered Khiva and a formidable group of Gcaleka in a deep and bushy ravine and backed off when they saw the odds were not favourable.

Khiva was on his way to the Ngqika Reserve north of Draaibosch and across the Kabousie River (a tributary of the Kei and joining it from a westerly direction). He had an important message to deliver to Sandile from Sarili. Sandile, too, had resisted efforts to drag him into the war but some unfortunate incidents had excited the mood of the Ngqika. In December, the old chief gathered his councillors and people and addressed them saying, 'How can I sit still when Rhili fights? If Rhili fights and bursts and is overpowered, then I too become nothing.'

His chief councillor Tyhali refused to contemplate joining the fight and took his followers to Mgwali to seek the protection of the resident magistrate. But Khiva and Sandile's son Matanzima fell on some Mfengu kraals, destroyed the hotel at Draaibosch and ambushed the postriders riding between Komga and the railhead at Kei Road.

Sandile and his Ngqika made for the near-impenetrable bush of the Tyityaba valley, another tributary running into the Kei from the west (there is a sign pointing to it on the main road between Komga and the modern bridge across the Kei). Here they were joined by Sarili and the Gcaleka who managed to evade Cunynghame as he swept eastwards towards the Mbashe. So, at last, the Gcaleka and the Ngqika were united and they were behind Cunynghame.

On 13 January, Colonel Richard Glynn with elements of the 24th and 88th Regiments, some police and marines, sallying forth from the camp at Ibeka, inflicted a sharp reverse on a large force of Xhosa at Nyumaga, between the Kei and Qolora rivers, which sent them scurrying back to the Tyityaba.

Sandile and Sarili were now in a difficult position. Supplies of food and particularly ammunition were desperately short. Sandile wanted to raid the Mfengu and seize cattle; Sarili wanted to overpower a supply camp and acquire much-needed military hardware. This was a far more dangerous enterprise but the latter, being paramount chief, prevailed.

Centane Hill lay between the Tyityaba valley and Ibeka camp. Cunynghame had deputed Captain Russell Upcher of the 24th Regiment to set up a forward supply base there when the invasion of Gcalekaland was planned. Upcher first formed a wagon laager, then had his men construct an earthwork fort surrounded by a deep ditch. He had with him two companies of the 1/24th Infantry Regiment, a handful of police, two guns and some Royal Marines with a rocket launcher – in all some 700 men. These were augmented by 300 Mfengu under the leadership of the famed fighter Veldman Bikitsha.

Mfengu scouts began to report, by the end of January, that the Xhosa were beginning to emerge from the Tyityaba and were massing in the nearby Mnyameni bush, so Colonel Glynn reinforced Upcher with a detachment of Carrington's horse.

Upcher had a line of rifle pits dug a little way in front of the existing fortifications. Both the fort and the rifle pits can clearly be seen to this day.

Slowly the Xhosa moved on the depot. Sarili addressed his men with power and passion. He knew that this was his last gamble.

Visibility was not good at dawn on 7 February 1878. This allowed the Xhosa to approach quite closely from the south before the Mfengu sentries descried them creeping though the mist and drizzle. Bugles immediately sounded the alarm and the 24th, issued with 70 rounds of ammunition apiece, took up their places in the rifle pits.

But the Xhosa did not charge and Upcher had to send out a skirmishing company of infantry to occupy a small hill to the north to try to lure them on, while Carrington probed the Mnyameni bush. This at last provoked the Xhosa who launched themselves in hot pursuit of Carrington's retreat. They reached the screen of Mfengu beyond the entrenchments, forcing them to retreat, too. The artillery shells and rockets failed to stop this charge, as brave as Major-General Pickett's at Gettysburg.

But then the infantrymen, concealed in the rifle pits, revealed themselves and sent a volley into the Xhosa that mowed them down like grass. Muzzle-loaders were no match for the new Martini-Henrys. The late last flower of Gcaleka warriordom, after 20 minutes, wilted. They were pursued by Carrington's horse, the mounted police and the Mfengu.

When the cavalry returned to the camp, breakfast was being served. But the British were in for a sharp shock.

To the west, Sandile's Ngqika, hitherto unengaged, appeared in threatening mode. Experienced fighters as they were they set their own trap, not dissimilar from the one they had just seen the British use. When a force of police and infantry descended on them, clearly as bait once more, they lured this bait further than they should have ventured. A hidden Ngqika force suddenly appeared in their rear and threatened to cut them off.

Upcher hastily sent reinforcements but these, too, got into difficulties and the fighting became hand-to-hand. The infantry on the left of the police, disciplined troops as they were, closed up the line and the desperate situation stabilised somewhat. But it was still touch-and-go.

At this moment the approach of a small detachment of police and mounted volunteers, encamped some miles away, was reported to Sandile. He did not know its strength. Had he heard of Napoleon's great defeat at Waterloo, where Blücher's timely intervention inflicted a crushing blow to one of history's greatest generals? He decided he would not be caught between two forces and so he led his warriors away to a conical hill out of reach of the guns and rockets.

He paused there for a while watching Upcher regroup inside the camp. The battle was over.

Upwards of 400 Xhosa died that day.

The Ngqika were not done. They retreated to the Kei valley, then, eluding their pursuers, flooded back into their beloved Amatole mountains from which they had been forcibly excluded years before. Many prominent leaders joined Sandile there, including Dukwana, the Christian son of Ntsikana. Some did not. Old Soga, father of Tiyo, refused to fight but did not join Tyhali. He went home to his village. When the Mfengu dark avengers inevitably came for him he asked only to be killed with his own assegai.

For months the Ngqika resisted the powerful army of General Thesiger, Lord Chelmsford. But at last, in early June, Sandile, of the crippled foot and handsome mien, was cornered and killed. Then the Ngqika finally submitted.

Centane Hill spelt the end of the Gcaleka as a fighting force. Sarili died in exile beyond the Mbashe River in 1895.

Pulleine, Carrington, Upcher and the 24th Regiment went off to Zululand where, on the night of 21 January 1879, they camped under the shadow of a hill called Isandlwana.

CHAPTER SEVENTEEN

THABA-MOOROSI

Mount Moorosi might seem one of the most remote places in the country to visit, and one of the most unlikely, since most people will never have heard of it and, even if they have, will not have any compelling reason to go there. But they are missing something and the area is easily accessible by a good tarmac road and it is our duty, perhaps, to give them some pretext, however minor or spurious, to spend a day on a trip there.

The most convenient bases are either Zastron or Mohale's Hoek (where there is a hotel, though I have never stayed in it). The dirt road from Zastron through the border post at Makhaleng Bridge is a good one and the nearer it approaches Lesotho the more dramatically do the layer upon layer of mountains unfold. Once in Lesotho (you need a passport to enter) the two hour drive to Mount Moorosi via Quthing contains scenes and places of no little interest.

A modern bridge spans the sometimes lazy, sometimes turbulent Senqu (Orange) River and this side of Quthing are three notable historic sites. The first you come to on the right-hand side is Masitise, one of the mission stations of the Paris Evangelical Missionary Society (PEMS). Do not take the road to the high school but wait for the second turn-off to the primary school. Drive up to the old church and ask at the house of the minister next door for the way to the cave. This is a short, easy walk up some rock steps.

Here, in 1866, Frédéric Ellenberger built into the overhanging rock his first mission house, which survives, in good condition, to this day. A house in a cave! Here, too, in one of the small adjunct rooms, Ellenberger set up the mission printing press.

The press had an extraordinary history. It had been brought to South Africa in 1835 by the PEMS missionary Samuel Rolland, but lay in its original container unused for six years at his mission station at Beersheba. Thereafter Rolland printed several of the earliest books in Sesotho on it (including the New Testament). In 1848 it was badly damaged in the most extraordinary way. The printing rollers were made of gelatine and glycerine, so tasty that one night some jackal made their way into the printshop through the door which had no lock, in order to sample the delicious offering. When they repeated this the printer stayed up one bright and moonlight night with his gun; but the shot exploded in the magazine, wounding his hand and preventing him from plying his trade from then on. In 1858 when war broke out between the Boers and the Basotho, the press was hastily moved from Beersheba, which was threatened by the Boers and subsequently shelled and sacked. The type was hastily loaded into sacks and was consequently horribly jumbled into the early equivalent of alphabet soup. On its way to its destination the wagon broke down (so heavy was the type) and a few of the sacks burst, spilling the type on the ground. Some passing Basotho warriors gathered it up and took it to the cave to make rifle bullets. The word made lead. Ellenberger immediately charged up the hill to rescue it and was nearly shot, being mistaken for an enemy. He was only saved by a Mosotho who was one of his converts and who recognised him. Ellenberger reassembled all the type like a massive jigsaw and the press ended up in a small room with a mountain for a roof.

Masitise is a must-see.

Within sight of the Protestant station, as though cocking a snook at it, are the bizarrely ostentatious twin spires of the Roman Catholic church of Vila Maria. Just outside Quthing (that typically vibrant, chaotic Basotho town), again on the right, is the Leloaleng Technical School, which is also worth a visit though the access road is atrocious and needs to be negotiated with care. Leloaleng was another creation of the Paris Mission, founded in 1880 as the first industrial school to train Basotho men in the crafts of masonry and carpentry. There are some fine buildings there, though in a state of benign neglect.

The 40 kilometres of road beyond Quthing to Mount Moorosi largely track the course of the Senqu, with its dark green waters, and magnificent prospects. Mount Moorosi is a small town of taxi touts and jivy music, old ladies hobbling painfully with sticks seemingly from nowhere to nowhere, donkey carts overloaded with straw and herdsmen pushing their stock through the crowded main street. All a far cry from the quietude of the middle nineteenth century.

Though this was shattered in 1879 when the ruthless Governor Frere and the priggish Cape Premier Sprigg, fearing a hostile confederation of African tribes, spurred on by the Zulu success at Isandlwana, decided to disarm them, starting with the Baphuthi under Chief Moorosi, holed up in his mountain fortress. They used as an excuse a minor transgression of Moorosi's son, Doda. A force of Cape Mounted Rifles (CMR), Cape Mounted Yeomanry, Mfengu levies and Basotho under Lerotholi (brother of the Basotho king) invaded the mountain and attacked it at night on 7 April. Even with artillery, they were repulsed with the loss of five killed and seventeen wounded (two Victoria Crosses – to Sergeant Scott and Private Brown – were awarded in the action). Subsequent actions resulted in failure and loss, including a massacre when the Baphuthi surprised a small camp at night, killed the sentries and assegaied the men as they struggled to get out from under the canvas of their tents. Despite parleys the mountain fortress still held out into October.

The town of Mount Moorosi is not the original site of Moorosi's stronghold. You need to travel further for that. Drive 1.3 kilometres beyond the Pep Stores. On the left of the road is Khabisa Wholesalers. If you look ahead and slightly to the left you will see Moorosi's mountain. Another 1.2 kilometres round the bend, you will get a closer view of it, and will be approximately where the attackers established their camp. If you look up you will see, from right to left, the significant features of the battle – the Neck, the Dip, the Saddle, the Lip and the mountain itself, with a large flat rock on its right hand (or southerly) side. The road carries on round the base of the mountain, which is on its right-hand side with the Senqu (or Orange) on the left, until it crosses a bridge over the Quthing River which joins the Senqu at this point. The mountain lies within the embrace of the Senqu and the Quthing. On the night of 19 November 1879, it was the scene of a dramatic battle.

There is a spirited eyewitness account (unpublished as far as I know) of the final campaign written within a couple of weeks of the events themselves. It is anonymous, though an accompanying sketch has a signature which might, with some difficulty, be deciphered as something like Mathew Mattisson. The author of the manuscript seems to have been an orderly or soldier attached to the medical party of the attacking force. We shall call him the Unknown Soldier.

The colonial army comprised some 1 400 men of whom the majority were men of the CMR led by captains Bourne and Montague and of the Herschel Native Levy under Commandant Alex Maclean and the resident magistrate of Herschel, Captain Hook (who, as far as is known, bore no resemblance to his namesake in *Peter Pan*). There were also 20 or so troopers of the Barkly Border Guards. Overall command fell to Colonel Zachary Bayly of the CMR. He was a man of about forty, of medium height, with fair hair and moustache, a bit stout, but every inch a soldier, who had made his name at the battle of Umzinzani in Gcalekaland. When he arrived at Mount Moorosi he sent the Yeomanry home, putting his faith in his favoured CMR.

While previous attacking forces had established themselves on the Saddle,

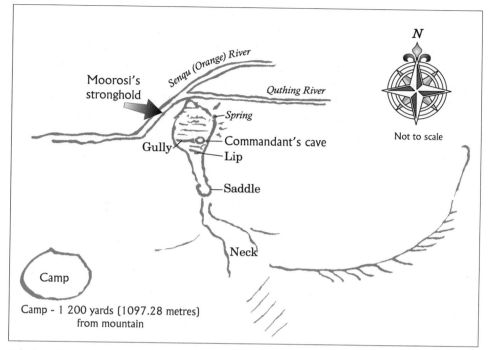

Thaba-Moorosi in November 1879 (traced from a contemporary sketch by an unknown soldier)

efforts to take the crown of the mountain had failed on 8 March and 5 June. These assaults were held back by heavy fire from the Baphuthi hidden behind 'sconces' (or schanzes) on the Lip. By mid-November Bayly was determined to make a full-scale final assault. There was some trepidation in the camp as heavy losses were anticipated. But Bayly had decided that frontal assaults had not worked. Because the Baphuthi were continuously resupplied, there must be secret ways up the seemingly sheer cliffs.

Bayly had three guns at his disposal — two 7-pounders and a 12-pounder. But the preliminaries to the assault began on the 12th with the long-expected arrival of a mortar which the Unknown Soldier suggested had 'probably come off a shelf in the Cape Town Museum' because it had the date 1806 stamped on it. But this 'venerable relic of bygone days' proved surprisingly effective despite the fact that the wrong fuses were sent with its 6 inch bombs.

The artillery troop of the CMR took it up to the ledges under the 'Neck' and tested a few shots at a distance of 1 100 yards. The Baphuthi, from the sconces, replied.

Next morning a fatigue party sallied forth to build a protective breastwork for the mortar 150 yards closer to the mountain. They were greeted with a hot fire, so some sharpshooters were sent out as a counter (though their old Sniders were not of the least use for sniping). The artillery guns also shelled the mountain with 'a very pretty fusillade' while the breastwork building was under way. The booming echo of an old elephant gun from the left-hand sconce above the Gully was almost as loud as the equivalent boom of the artillery guns. This gun was a percussion gun of enormous rifled bore made by E.M. Reilly of Oxford Street – such a gun as Bain or Anderson would have used to shoot elephants 25 years before. Now it was in the hands of Letuka, one of the Baphuthi defenders! Nevertheless, none of the builders was hurt and the mortar was soon brought into action, peppering all parts of the mountain.

Then, on 18 November, the second part of Bayly's strategy was fulfilled with the arrival from Aliwal North of some scaling ladders. There was a temporary setback, however, when it was discovered they were 'the most preciously idiotic productions conceivable for human ingenuity to have framed for such a purpose'. They were 3 feet wide, 12 feet high, with sides of deal an inch and a half thick, and actually broke in two with the weight of only one man! So Captain Bourne cut them down the middle (to reduce them to normal width), and killed four bullocks to make rawhide reins which he lashed round the sides and rungs. When the sun quickly hardened the rawhide, he had serviceable ladders.

Late that evening the colonel and a few of the officers scouted round the mountain and part way up the Gully without being fired upon. From this recce the plan of attack was formulated. The troops were to deploy the following night (19th) in preparation for its commencement early on the morning of the 20th.

The plan involved five separate incursions. On the north-east side of the mountain is a projecting rock with a spring below it: a hundred yards beyond it was a rough, difficult path, defended by strong sconces on top, leading to the summit of the mountain. Commandant Maclean was to try to reach the top via this path. More to the east of the mountain Bourne, with 200 men of the CMR, and Montague were to lead two parties which, faced with an almost perpendicular krantz, were to try to scale it with ladders.

Further to the south Captain Mullenbeck with the Border Guard and some 40 'Fingoes' (Mfengu) was to await the signal to rush over the top of the ridge between the Saddle and the Lip, a distance of some 350 yards fully exposed to the fire of the Baphuthi in the sconces 50 yards from the Lip of the southern end of the mountain. Finally, the fifth group, the Native Levy of 'Tambookies' (Tembu) under captains Hook, Peacock and Parker, was to make a feint up the Gully on the west side, no doubt to distract the defenders.

So, on the designated night, the whole force waited on tenterhooks as they watched the young moon's progress downwards in the sky until it vanished behind the high mountains to the north-west across the Senqu.

Finally, when it died at midnight, the small medical party, including Drs Hartley and Cumming and the Unknown Soldier, with stretchers and surgical instruments, headed for the Dip between the Neck and the Saddle. In doing so they saw a dark mass of men drawn up in silence under arms. When they reached the Dip, 700 yards from the headquarters camp, a long black file came, by the upper path, down over the Neck (making almost no noise as the colonel had ordered the soldiers to put woollen socks over their feet), paused for a few moments near the Dip and then went round the base of the Saddle, disappearing out of sight along the south-east slope of the mountain.

The Unknown Soldier and his companions stayed behind, lying under a stone wall which ran from the Neck across the Dip and part way up to the Saddle.

The sky was clear and cloudless and by no means pitch dark. So hushed was the air that it seemed impossible to move without being noticed by the Baphuthi. So still was the night that the sound of the Senqu rushing over the drift more than a mile to the west could be heard, as could the soft flutter of the wings of night birds passing overhead. The Saddle rose abruptly above the men in the Dip for 200 feet, its black mass resembling a perfect pyramid which completely blotted out the whole of the mountain from them. A dog up above barked intermittently but otherwise there was no sound. The Unknown Soldier almost dozed off on a stretcher.

Later an orderly brought a letter from the camp for the surgeon Dr Hartley. There had been an awkward incident at the Gully. At the critical moment, when the attack was about to begin, Hook suspected the Tambookies either of cowardice or treachery so he sent them back to camp where the guards, too few to disarm them, turned their guns on them ready to fire if a breakout was attempted.

The Unknown Soldier and the game old Dr Cumming struggled up the almost perpendicular approach to the Saddle, rough even in broad daylight, painfully slow in the dark. Near the top the Unknown Soldier heard voices coming from below and asked one of his bearers to interpret. The talkers were Tambookies who were saying that they did not want to 'schit' (shoot) any 'Maphuti' but would get up into the 'schantzes' and 'help schit the white man'!

When the 40 Fingoes and the Border Guard under Captain Mullenbeck (as well as the Unknown Soldier) arrived in front of the left-hand sconce on the Saddle, the prospect of a successful attack seemed dubious and the men manifested distinct reluctance to do so, and all of them chattered away like a flock of babblers in a bush until they were hushed by the old medical man.

There was hardly 15 minutes of darkness left when the flash and report of a gun came from the extreme north-east point of rock above the Spring. It was followed by several more flashes showing that the men on the ladders had been discovered. On the Saddle Dr Cumming, with his white helmet off, leant far over the breastwork, straining to make out with his field glasses what was happening on those dimly seen grey patches of rock half-a-mile off. Mullenbeck advanced his Border Guards 50 yards.

What ensued was chaotically episodic for the Unknown Soldier. There was a burst of shooting, some cheering, three notes on a bugle. The CMR had evidently gained the top. So, too, it seemed had Maclean and his Fingoes on the northern side. Daylight came and the flashes advanced rapidly along the top of the mountain. A shell from the camp fired into the sconces was the signal for the

Border Guards to rush the 350 intervening yards to the Lip. Mullenbeck drew his sword and waved it dramatically above his head, his steel scabbard ringing sharply against the stones as he yelled 'Border Guards, advance!' His men showed no inclination to follow him as bullets came flying at them. It was apparent that if they could not get to the Lip they certainly would not rush the sconces 50 yards above it.

A cheer rose from the summit and there was the unexpected sight of a white man rushing, flag in hand, to the big rock (informally known as the 'Comb'). It was Lieutenant Sprenger of the CMR, the first man up the ladder. He had alighted on a slender ledge which could only hold a few men. One of the brave defenders had thrust at him with an assegai, piercing his felt hat, only to be shot dead instantly, his body falling over the edge of the krantz onto the men below waiting to scale the ladder. These men were also subjected to a weird experience: the Baphuthi for some unknown reason had stretched out frameworks of bone and dried-out skin of dead animals and one of these skin-and-bone dead horses was pitched onto these attackers below to their singular consternation since, in the darkness, they had no idea what rattled so!

Once 20 men had gained the summit they rushed after Sprenger with fixed bayonets over the top to the south end till they reached the Comb. Maclean with his Fingoes had also reached the top. The Baphuti in the sconces below now turned their fire on the big rock but a murderous fire from the Sniders poured down on them, killing 25 of them very quickly. Those who survived fled down the mountainside towards the Quthing River pursued by the Fingoes or flung themselves over the steep krantzes towards the Senqu.

In the wake of the assault troops the Unknown Soldier was impressed by the courage of those CMR troopers who had scaled the ladders. Had they done it in the daytime they would have turned giddy – as many had never climbed a ladder in their lives! So it was better that they had done it at night.

What was clear – and lucky – was that the Baphuthi were taken by surprise. They had a complacent sense of security that the sheer cliffs could not be scaled. Also, said prisoners afterwards, the continuous random fire of the mortar had prevented sleep so effectively the night before that many behind the sconces had fallen asleep at their posts. Bayly's plan had worked in this respect.

The victors on top found a small village and forthwith set fire to the thatch so that the flames and the drifting smoke in the early morning proclaimed all over Basutoland that Moorosi's mountain was taken. Through the haze a crowd of CMR clustered round the Union Jack planted on the Comb.

It was soon put to good use a few minutes afterwards. Sergeant Neville draped it over two prisoners whom 'the Fingoes showed an inclination to despatch'. But the Unknown Soldier was deeply disturbed when Lieutenant McMullen of the CMR, interrogating a prisoner who refused to reveal the whereabouts of Moorosi, put a carbine to his chin and blew his head to pieces. He was pleased when Sergeant Nelson, also of the CMR and in an identical situation, saved his trembling but nevertheless mute captive by declaring that he admired his bravery.

When Colonel Bayly arrived on the Comb, dapper in his deerstalking hat, cardigan jacket and yellow riding cords tucked into his socks and with a stake in his hand, he was loudly cheered because his careful planning meant that there had been, contrary to expectations, no fatalities on the colonial side. He said a few suitable words and promoted Sprenger to captain on the spot. (The gallant Sprenger was killed at Wepener in the Anglo-Boer War 20 years later.)

Everybody then devoted themselves to sightseeing, exploring the stronghold which had been so long defiant. Behind Moorosi's dwelling a huge watchdog showed grimly its teeth even in death. The Unknown Soldier lamented the destruction of the beautiful and varied prehistoric earthenware pots in the flames. He was also taken aback by the lowness of the sconces, which could have easily been flattened by the artillery had they known that they were not five feet high as thought but only half that size.

One of the searchers picked up Moorosi's Bible, a large quarto in brown leather binding. He began diligently riffling through the pages of the Scriptures looking for the cheque of 200 pounds Moorosi was known to have possessed at the outbreak of hostilities.

The Unknown Soldier took the time to appreciate the wild and beautiful panorama with the light falling on 'the roseate coloured bergs' and went to gaze down the magnificent precipice on the western side to the Orange River below, which 'reflected every tint of the sky overhead like a glass'. Others explored the hollow ledges like galleries in the face of the rocky sides.

By 7 a.m. fatigue drove most of the men, who had not slept that night, to move towards camp. But as the Unknown Soldier passed down into the left-hand sconce above the Gully which Letuka with his elephant gun had often frequented, his attention was drawn by a commotion caused by Private Whitehead on a ledge below. He could not hear what Whitehead was gabbling but the private was pointing to this kepi, the peak of which had been ripped up by a bullet put there by some lurking fugitive on the ledge.

It appeared that Whitehead had found Moorosi in a sort of cave or hollow. It had been Doda who had fired on Whitehead. He began flinging stones out of the cave at the men who now approached it and then fired a shot which hit Private Schwach of the CMR in the thigh. Doda, kneeling behind a flowering bush at the entrance to the cave and armed with a rifle and carbine, fired at every man who eased himself around the corner of the narrow ledge in an attempt to flush those in hiding. Eventually Doda daringly escaped along the ledge on which even a mountain goat would have difficulty in negotiating a foothold and threw himself over a cliff towards the Orange River below.

But Moorosi himself was found with two fatal wounds, one on the left side of his neck, the other in the ribs. Also found in one of the lower sconces, was the corpse of an albino, which explained the rumour that there was a white man on the mountain. He was 'of a distinctly negro type' but had white skin and ginger-coloured hair. His name was 'Blesskop' and he was reputed to be the marksman who had caused so much havoc in earlier assaults (the CMR called him Jonas and long dreamt of what they would do to him if they caught him). Of the estimated 170 defenders about 80 were killed and 14 taken prisoner. The rest escaped down the precipices or died in the river. Doda ('the last of the Moorosi race') was found in a cave seven miles away, dying from a shattered thigh caused by his 'tremendous leap for liberty' down the side of the mountain. It was a mystery to his finders how he could have made it so far with such an injury.

The Tambookie mutineers were disarmed and five of the principal sergeants and headmen were given 25 'very mild' lashes with a stirrup leather. The last to be flogged was asked why they had behaved so treacherously. 'Because,' he replied, 'the Sergeants and Corporals led us astray.' Seventy of the rebels were stripped naked and booted out of camp, amidst hearty laughter, and harried into the veld by the loyal levies. The CMR hastened away across the Drakensberg to Kokstad since there was trouble in Pondoland which needed their attention. Said the military historian of the campaign (Major G. Tylden): 'The lesson of the final assault was plain enough: given a commander who understood the use of artillery, no hill fortress could hope to hold out against white troops.' Bayly had only lost three of his men, thereby winning the loyalty of his troops.

Moroosi's body was brought into the camp. His age was estimated at 68 and he was still muscular. He was no more than five foot eight in height and he had 'a fine head full of crafty intelligence'. In death his features preserved 'a curious life-like expression, the half-opened eyes and compressed thin lips gave a curious disdainful expression to his face'.

What the Unknown Soldier did not say was that he was then decapitated and his head dispatched to King William's Town (though the PEMS missionary Adolphe Mabille remonstrated so volubly that it was later returned to his home for burial).

This barbarous act was not forgotten by the Basotho. In the tragically unnecessary and vicious Gun War that broke out in the following year some Batlokoa loyal to Paramount Chief Letsie of the Basotho killed and decapitated the magistrate John Austen and sent his head to the paramount chief.

So the grim game in which the British and Basotho went head-to-head ended in a draw, 1–1.

CHAPTER EIGHTEEN

KANONEILAND

Rogues, ruffians, renegades. Outcasts and outlaws. Rustlers and rievers. Scoundrels, and dogs and black sheep. These are the terms to describe the denizens of South Africa's Wild West in the nineteenth and early twentieth centuries.

Cupido Pofadder, Jonker Afrikander, Klaas Springbok, Dirk Vilander, Carel Ruiters, Piet Rooy, Jan Kivido, Scotty Smith, Manie Maritz – these are the names evocative of the same area and the same period.

There is a great book to be written about the Northern Cape. If you dig hard enough you can put together a picture of its historical richness through publications like the sadly now-out-of-print *Forgotten Frontiersmen* by Alf Wannenburgh, *The Colonisation of the Southern Tswana* by Kevin Shillington and *Lodges in the Wilderness* and *Between Sun and Sand* by W.C. Scully, who was magistrate at Springbokfontein in the 1890s. There is a biography of Scotty Smith by F.C. Metrowich and on the early inhabitants there is J.A. Engelbrecht's *The Korana* and an obscurely published but excellent micro-study by Teresa Strauss called *War Along the Orange*. At local libraries and municipalities you can sometimes find local publications such as Maria de Beer's *Keimos en Omgewing*. In particualr there is a fine chapter on the Korana in Nigel Penn's *Rogues, Rebels, and*

Runaways. One must always bear in mind in what happens that Korana identity (as for other such groups right up to the present day) is as elusive as shifting sands and sometimes gravitates to where material or spiritual advantage presents itself.

It's surprising that more people driving between Johannesburg and Cape Town, or vice versa, don't for a change take a couple of days extra and go the longer way round through fascinating countryside via the West Coast and Northern Cape. Along the Orange – the Green Kalahari – the countryside nowadays is lush and relaxing, the grapes in season are sweet, and the local boys will tell you that their dates are the best in South Africa and that their girls aren't bad either.

It hasn't always been like that. Before the Second World War life was the epitome of hardscrabble. The first 52 settlers, without proper permission, encroached on Kanoneiland in 1928. One of the farmers there, Hennie Steyn, told me that '*Die hele ding was bebos*'. And that trees cut down had to be buried on the spot. For years they scratched a bare living. Success eventually came but the secret was in moving off the islands to the richer alluvial ground on the banks. At Kakamas (where irrigation was to give the lie to its name which means 'poor pasturage' and was eventually to produce the first peach variant – the Kakamas peach – suitable for canning in this country) the first settlement (begun in 1898 under the auspices of the Dutch Reformed Church) struggled hard to dig canals and clear the bush and really only began to prosper after the introduction of sultanas in 1928.

In the 1820s the Gariep (or Orange) was visited by the traveller Thompson who described the people he found there: 'The Korannas are a race of pure Hottentots, who have attached themselves to the vicinage of the great River.' Some believed that their forefathers came from the Cape Peninsula. These semi-nomadic pastoralists spoke a dialect of Khoikhoi, lived off the products of their cattle, hunted in the Kalahari and gathered *tsama* melons and wild potatoes from the veld. During the following decades, skilled horsemen, they ranged the plains to the east, even to Lesotho, and some of their fighters fought alongside the Barolong at Viervoet in 1851.

After the middle of the nineteenth century there were several bands of Korana living along the Gariep. At Kheis on the edge of Griqualand West were the Bovenstanders. Next to them to the west were the Springbokke under Klaas Springbok, the most populous of the groups, numbering a few thousand only. Then there were the Katte (Katse), or Cat Korana, who lived around the drift then known as Olyvenhout's Drift (later known as Upington). Their chief was Klaas Lukas, with Gert Perkat and Klaas Papier occupying the contiguous territory, and further down-river were Jan Kivido and Piet Rooy. Between them they could raise perhaps a few hundred fighting men and no more. The most westerly Korana group under their chief Cupido Pofadder occupied the area around Kakamas. To the west and north of them were non-Korana groups, the Afrikaners and the Bondelswarts (with their main town at Warmbad).

Korana society was cattle-based and loosely organised. Wealth and status depended to a large extent on cattle and access to pasture and hunting grounds were critical to their existence. Much of their surplus time and energy was spent cattle-raiding and thieving between themselves and from any other groups who happened to be within reach. The feud was as popular amongst them as it was between the Shepardsons and the Grangerfords in *Huckleberry Finn*. They were expert horsemen, marksmen, trackers and guerrilla fighters and ambushes must have been the fatal end of many a less-skilled opponent – individual shootouts unrecorded in time. Above all they used the terrain they occupied to their maximum advantage.

The banks of the Gariep were at the time dense with foliage, bush and reeds. Furthermore, dotted along the river itself were numerous islands, small and big, difficult to reach especially in the summer months because of the flooding, and almost impenetrable except to those who knew the secret pathways. They represented a defensive fallback of the kind the Somerset marches around Athelney provided King Alfred against the Vikings in the ninth century.

Their domination of the area should not be sentimentalised. It was not achieved entirely bloodlessly. There were several wars with the Bushmen and a bloody defeat of the latter occurred at Koukonop's Drift. In the eighteenth century they suffered, too, from a devastating smallpox epidemic.

But by the 1860s their *liebensraum* was being squeezed, by coloureds and Xhosa to the south-east around Prieska, and by the Trekboers moving north, first on the Hartebeeste River and soon after to the Orange. What exacerbated the problem was that the colonial government preferred to pretend that the northern border did not exist. The Korana's pasture and their water were now a matter of frequent contestation. But at the same time a new opportunity was presented for their special skills and appetites – rustling parties raiding south in search of trekboer cattle.

The external pressure helped cause an internal implosion in 1867 when the Springbokke (under Klaas Springbok and supported by Gert Ruiters, Piet Rooy, Jan Kivido and Cupido Pofadder) squared up against the Katte (under Klaas Lukas backed by Perkat, Papier and the Bovenstanders). The latter defeated the Springbokke and drove them north into the Kalahari. Rooy and Kivido became more powerful, though Lukas and Pofadder, now in alliance, put them in their place. Rooy harboured a simmering grievance against the colonial authorities after being subjected to 3 months' hard labour and 12 lashes for an offence for which he might well have been innocent.

Coloureds moving north across the Orange ... Damara (Herero) moving south into Namaqualand ... drought throwing a blanket over everything – an explosive mixture.

In 1868 a desultory war broke out between the Colony and elements of the Korana and Rooy. A small force of mounted men, gathered into the Northern Border Police under Special Magistrate Maximillian Jackson, achieved little and in fact had a setback in a skirmish at De Tuin where they lost two men and all their weapons and equipment. In June 1869 a combined force of Frontier Armed and

Mounted Police, Northern Border Police and some coloureds and blacks (about 250 men) under Sir Walter Currie besieged and tried to starve out the Korana from their islands with only minimal success.

Real success only came with a cunning change of policy. Set a Korana to catch a Korana was the new approach. Klaas Lukas was recruited to talk to Piet Rooy, Jan Kivido and Carel Ruiters and try to persuade them to talk peace with the Colony. The Korana met on 26 October and Lukas took the three obdurate leaders prisoner. Ruiters escaped but Rooy and Kivido were handed over to Jackson at Kenhardt. The remnants of the resisters were remorselessly hunted and ground down and in February 1870 Ruiters was finally recaptured.

As a reward for defending their own territory Piet Rooy, Jan Kivido and Carel Ruiters were packed off to Robben Island.

One tragic side effect of this first Korana War was the virtual extinction of at least one band of Bushmen in the area. After sporadic attacks they made on the oasis of Pella, raiding the gardens and running off with cattle, they were overhauled by an alliance of colonists and Bondelswarts and the entire band, including women and children, were cut down, many killed, many wounded.

In 1870, in terms of a treaty conducted with Jackson, the Korana were confined to the northern bank of the Orange – Klaas Lukas counted his territory to extend from the edge of Griqualand West to Currie's Camp, and Pofadder from Currie's Camp to the Augrabies Falls. The islands on the river were excluded from their authority. In 1871 a mission station was established at Olyvenhoutsdrift by a man who was to have an enormous influence on the area – Revd. Christiaan Schröder.

But throughout the decade of the 1870s the world began to close in on the Korana. Ngqika moved into Prieska as squatters, coloureds put pressure on land, drought and loss of grazing and hunting pushed many into outlawry, and alcohol sapped their resolve and their capabilities.

Their final tragedy in 1878 to 1879 should not be seen in isolation. In the eastern Cape the Gcaleka and the Ngqika were making their last stand and in 1878 there was a serious rebellion in Griqualand West, the ripple effect of which spread westwards downriver. Two of the leaders of this rebellion, Jan Pienaar and Donker Malgas, fled to the protction of Klaas Lukas and they were to fight alongside the Korana in the conflict that was about to erupt. Was there some kind of concerted conspiracy between the Xhosa, the Griqua, Sesotho-speakers and the Korana? The best commentator on this point, Teresa Strauss, suggests that 'there is no positive evidence of any long-planned and calculated scheme for warfare against the colony' and that therefore the jury is still out on the verdict. But there is no doubt that there was some contact and a considerable knock-on effect.

In June 1878 the campaign aimed at the Griqua was launched against the Langeberg and Koegas. The 'rebels' were defeated in fierce fighting but not overrun. Many retreated westwards.

The colonial forces under Major R. Nesbitt believed they had broken the rebellion when they captured a large group of Xhosa in July.

However, the former friends of the Colony were now so disaffected that they had moved onto the islands and fought a desultory war against a weak and chronically disorganised colonial campaign. In October an over-estimation of Korana strength led to an ignominious retreat to Kenhardt by Nesbitt.

At the end of October a particularly horrible incident occurred at Luisdraai near Koegas. A party of Korana and Bushmen were attacked by a patrol and 46 were killed, including 10 women and children. The patrol commander, J.A. van Niekerk, gave an unconvincing account of the events and the episode became known as 'the Koegas Atrocities'.

In December Nesbitt again put together a strong force but, without a boat, was unable to cross the great river which was in flood. He could not therefore get to Lukas's strongholds and consequently resigned his position. Jackson returned to take his place. He found the colonial forces – an unhappy mixture of burghers, Bastards and Khoikhoi – demoralised and in total disarray. He therefore had to spend months reorganising and recruiting. He also set about building a boat.

He planned to cross the river, attack Lukas from the northern bank and drive him eastwards where he would be caught in a trap between his own detachment and a force from Griqualand West. He received a setback, however, when his erstwhile ally Klaas Pofadder (the successor to Cupido) followed the example of his men who had deserted to Lukas. And before he could launch his attack the arrival of Thomas Upington, the attorney-general of the Cape, to enquire into general mismanagement of the district, led him to resign in protest. He was replaced by Captain McTaggart.

The good captain promptly put Jackson's scheme into effect and crossed the river on 6 April. In the days that followed, backed by artillery, he systematically cleared the islands of Keveis, Skanskop, Kivido, Melkstroom and the main island. By the end of April many of the defenders had been killed and over 400 captured, though none of the leaders (Lukas, Pienaar, Malgas, Pofadder) were amongst them. Displaced, they simply shifted elsewhere.

The ins-and-outs of this scrappy campaign, which involved much precarious fighting in the thickets of the islands, need not detain us. They can be read about in Teresa Strauss's definitive account of the war. Suffice it to say that the Afrikaners joined the 'rebellion', though they and a large part of the followers of Pofadder were surrounded and disarmed in a surprise attack on their mountain camp below the Augrabies Falls. Most significant was the fact that the Bondelswarts, despite huge sympathy for the Korana, allied themselves with the Colony and helped in this ambush.

Lukas, Pofadder and Malgas retreated northwards into the waterless wastes of the Kalahari desert. They were pursued there by colonial posses under Captain Maclean who had succeeded McTaggart.

On 2 July Pofadder was captured. On 19 July Maclean took a force of 119 men to a waterhole 65 miles from Kakamas. In a dawn attack Malgas and 11 of his followers were killed. Lukas escaped but was captured later in the year. He died on Robben Island in January the following year.

The remaining leaders, also imprisoned on the Island, were released in 1883. They were old and broken and they could no longer revive the equally broken Korana societies.

A reminder of the original Korana can be found in the facial features and no doubt some of the genes of the local people but the language has disappeared forever. Only a few place names bring them strongly back to mind.

Unless one digs really deeply into the documents and newspapers of the time there are no specific magnets that necessarily draw one to individual spots. I prefer to drive and walk around the beautiful and fascinating area contemplating the whole campaign in general and imagining individual stories which may have happened anywhere along the river in the reeds and in the bush (of which there is relatively little left). Above all the river dominates the terrain and one's thoughts about tactics and strategy must give it a central role.

But if one wants or needs a specific place to visit then no doubt the place to go is South Africa's largest island, Kanoneiland. The origin of the name is disputed: one version is that the island was so bombarded with artillery that the name, as it were, stuck. The more romantic story has it that the defenders of the island under Klaas Pofadder in 1878, no doubt as a desperate answer to their chief tormentors, cut down a large quiver tree, hollowed it out in the form of a cannon and stuffed it with gunpowder. They must have loaded it with a cannonball which had come their way. When they lit the powder the improvised cannon exploded, killing six of the Korana in the vicinity.

Whichever version is correct the name of the island is a lasting memorial of the second Korana War and is symbolic of the kind of impersonal and superior forces which swept the Korana away like driftwood in a summer flood.

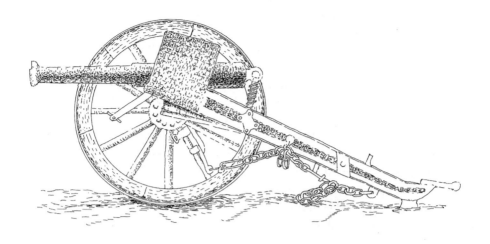

CHAPTER NINETEEN

POTCHEFSTROOM

Something heavy, something small; something precious, something tall.

This, in December 1880, was the formula not for a conjugal celebration but rather a messy divorce.

As with the unravelling of many a marriage this one began with a minor spat – involving Pieter Bezuidenhout (what was it with those Bezuidenhouts?).

The British authorities claimed Bezuidenhout owed some outstanding taxes: he, in turn, said his obligation was less than half of their demand and took his case before Andries Goetz, landdrost of Potchefstroom, who, in his turn, returned it to Pretoria, where the issue was handled with less than tact. Goetz ordered the impounding and sale of Bezuidenhout's wagon. On the day of the sale, 11 November, in Potchefstroom, the wagon was liberated by Piet Cronje and 100 armed and mounted Boers.

The administrator of the Transvaal, Colonel Owen Lanyon, immediately ordered Captain Pieter Raaff to Potchefstroom (the old capital of the early Republic) as field cornet to ensure order. Colonel Bellairs, military commander in Pretoria, also sent a military force under Major Charles Thornhill of the Royal Artillery to back up Raaff and Major Marshall Clarke, special commissioner in the town, overseeing all civil and policing matters.

Of course, the Bezuidenhout incident was not the main cause of the subsequent war. Ever since Theophilus Shepstone had annexed the Transvaal to Britain in April 1877 and the Union Jack had been run up by his young assistant, Rider Haggard, a large number of Boers (men like Paul Kruger, Piet Joubert and Piet Cronje) yearned for the restoration of an independent republic. The attempt by Raaff, intensely disliked by the Boers for his 'betrayal' of his volk, to arrest Cronje and others simply gave them further justification for action.

A large assembly at Paardekraal (west of present-day Johannesburg and a natural defensive position, within a day's ride of Pretoria) elected a triumvirate of Kruger, Joubert and Marthinus Wessels Pretorius (who had his house in Potchefstroom) and declared its independence. Each Boer present added a stone to a cairn indicating his pledge of loyalty. Raaff had two of his men at the gathering – one was a volunteer constable, Hans van der Linden, and the other a tailor with some medical knowledge, Christian Woite. On 15 December 1880 Woite's son brought a letter to Clarke in Potchefstroom giving details of the Paardekraal gathering, a letter which Clarke put in his pocket.

Pretoria did not seem unduly concerned with the unfolding situation but Bellairs did want his senior artillery officer back in the capital so he sent Colonel Richard Winsloe of the 21st Regiment, Royal Scots Fusiliers, to replace Thornhill. Winsloe covered the 160 kilometres in 48 hours and arrived in Potchefstroom on the afternoon of Sunday, 12 December. He was thrilled to have his own independent command which 'does not often come in the way of us soldiers'. The force he found consisted of 45 men of the right division of 'N' Battery, 5th Brigade, Royal Artillery (under the command of Thornhill and lieutenant Henry Rundle); 25 mounted infantry of the 2/21st Royal Scots Fusiliers; and two infantry companies of the same regiment (Captain Alexander Falls, lieutenants Peter Browne, Charles Lindsell, Kenneth Lean and James Dalrymple-Hay). There were also some commissariat and medical staff. Head of the former staff was Walter Dunne, who had played a prominent part in the defence of Rorke's Drift the year before and had been recommended for a VC for his role in the repulse of the Zulu there. In charge of the seven-man medical detachment was Surgeon Kenneth Wallis.

Amongst the artillerymen was 17-year-old Trumpeter Nicholas Martin and Driver Elias Tucker who had been among the very few to escape from Isandlwana where the other half of the N/5 Battery was massacred. Two others, Drivers Thomas Lewis and Abraham Evans, had also been at Rorke's Drift.

The garrison Winsloe took over numbered 213 men. The artillery was in possession of two nine-pounder guns.

Just outside the town Thornhill had begun to construct a regulation fort but a lack of urgency meant that little progress had been made on it, and it was not fully stocked with either food or ammunition. Winsloe accelerated its construction but inherited a badly sited installation. It was about 400 yards from the houses on the edge of the town with their walls and gardens. A water-furrow ran through the open ground between the fort and the town but its source, the Willows, was 1 200 yards from the fort. So the security of the water supply was decidedly dodgy.

Winsloe was not, however, unduly perturbed. The night he arrived he 'went down the town' after dinner to the fine house of Oskar Wilhelm Alric Forssman, a Swede of aristocratic birth. Chevalier Forssman was consul-general for Portugal (his title came from being made a knight commander of the Order of Jesus Christ). He had been granted a concession for supplying gunpowder, lead and stationery to the Transvaal Republic and had been a member of the Legislative Assembly under the British administration. A great landowner, with over 100 farms and numerous urban properties, he was almost certainly the richest man in the Transvaal.

That night friends dropped in and the gentlemen's talk was that there would be fighting ere long but the officers gave this no credit. The ladies of the family sang to them and walks on the verandah in the moonlight were taken. A ball was arranged in two nights' time. And the following two nights were spent in similar fashion, with whist and singing and romantic walks. Winsloe himself seems to have been particularly enchanted with Emelia, the vivacious 21-year-old daughter of the Chevalier and wife of Dr Charles Sketchley, the district surgeon. (So much so, that General Redvers Buller, in his subsequent report of the episode, said that Winsloe was 'so taken up with Mrs Sketchley that he let all things slide'.)

But, on the 15th December, as Thornhill was about to leave for Pretoria in the post-cart he observed a large force of armed Boers (there were about 800 of them) ride into town, so he galloped back to the fort to give warning of the development. The fort was immediately put on a war-footing. Two other points were also fortified: about 350 yards to the east of the fort and on the edge of the town but standing in open ground was a square building with walls 20 foot high which was the gaol; 300 yards east of that, on Market Square (close to where the modern post office now is), was the landdrost's office, symbol of governmental authority and the place where the new telegraph was housed. The fort itself was on a gentle slope (from west to east) and the guns were placed in shallow gun-pits on the north-east corner of the fort outside the walls and facing the vulnerable cemetery some 360 yards to the north-east.

These three strongpoints were to be defended. The parapets of the earthen fort, by no means complete nor impregnable, were hastily heightened and strengthened with mealie sacks and crates of tinned beef from the reserve supply. That evening some fifty citizen refugees from the town sought sanctuary at the fort. These included Chevalier Forssman, his wife, young son and seven daughters (among whom were Emelia Sketchley and Mrs Katherine Palmer), and two teachers, the Misses Malan and Wart.

While he could not '   was 'embarrassed by the presence of ladies and children'. To   with the   'ansport drivers and the regular soldiers the inhabitant   fort eve   reached more than 300. And the structure was no more

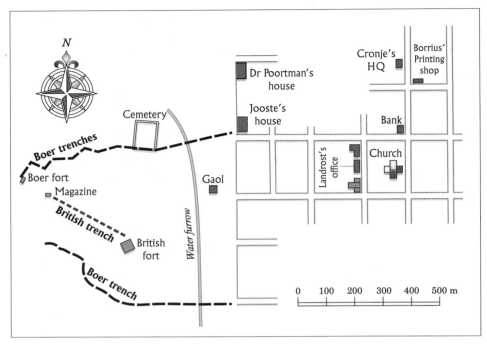

The seige at Potchefstroom (based on a drawing by Gert van den Bergh)

The landdrost's office was garrisoned by Captain Falls and 24 soldiers (they were joined by Major Clarke and Raaff and a few of his men, who were scattered in nearby buildings to provide cover fire). Lieutenant Dalrymple-Hay and a detachment of Fusiliers guarded the gaol. But the Boers held the rest of the town (including the Royal Hotel, the Forssman house and Goetz's residence). There were some British loyalist families in town but the majority of the town residents were firmly in the Boer camp. Cronje's first objective was to get the Proclamation of Independence printed by the sympathetic printer (his name was Borrius) in town.

But at 9 o'clock on 16 December (a day of real significance for the Boers as the anniversary of the battle of Blood River in 1838 and the covenant that was made then), while the British breakfasted in little groups outside the fort, guns to hand, a small party of Boers rode slowly past, about 150 yards away, with their rifles at the 'carry'. At this the Mounted Infantry, whose horses were already saddled, were sent to warn them that the Boer patrol was not permitted so close to their own lines.

The Boers turned and rode off at a trot, followed into town by the British. At this the first firing of the war began. If both sides are to be believed they were not the first to shoot — and therefore no fighting could ever have started! The truth is that it does not really matter — both sides were able and willing to begin the battle. There was only one casualty — the Boer leader of the patrol, Commandant Robbertse, was wounded in the arm — but as historian Ian Bennett has written, this was 'a day which arguably changed the course of the history of South Africa'.

The first shots did unleash a general fight, however. The mounted infantry retired under the protection of the artillery guns which were now surrounded by a few mealie sacks. The horses were hidden in the fort's ditch. The Boers poured into the market square, occupying the houses and gardens and directing fire into the landdrost's office and into the gaol. Their right flank pushed into the cemetery, positioning themselves behind its walls, but the fire of the British artillery became too hot for them and they were driven back. After 20 minutes the attack ran out of steam. Despite the heated affray, casualties were slight (the Boers never revealed their losses). But Captain Falls in the landdrost's office was dead, shot through the heart, having time only to say, 'Oh God,' before he died.

Next day firing was continuous but had little effect on either side. But on 18 December the Boers poured fire from their Wesley-Richards rifles into the landdrost's office turning it, despite its sandbagged windows, into a sieve. Captain Falls was actually killed in the passage and later buried in the garden of the Standard Bank (I wonder if my own bank manager would be so accommodating). Then one of the most courageous of the Boers, Jan Kock, climbed on the roof of an adjacent stable, and flung home-made spears made out of long reeds and wound round with burning turpentine-soaked rags onto the thatched roof of the office. When the thatch began to burn Clarke knew further resistance was futile and he gave Raaff permission to surrender. Clarke himself was to be treated with civility by Cronje but the despised Raaff was kept manacled for many weeks. Winsloe had never wanted to defend the landdrost's office – it was too cut off from the fort – but was constrained to do so because of prior decisions from his predecessor who was concerned about its symbolic significance.

Now he chose to abandon the gaol, too (since its sun-dried bricks provided little protection), and arranged to do it at night when it was slightly safer. Dalrymple-Hay and his men were to wait for the signal – a lighted lantern placed on the fort's parapet. He had already lost a man and had three wounded to bring with him. Surprisingly, the Boers did not detect the move and it was carried out without mishap. Next day the Boers were taken aback when they attacked an empty building.

Over the next few days Winsloe was assailed by a new problem, the most critical of all – access to water. A well was dug but it hit rock. They dug through 16 feet of rock to a depth of 30 feet before they found water but it was half mud and of negligible quantity. For three nights they escorted water-carts to the fountainhead at the Willows 1 200 yards away. But then this exercise became too dangerous and the water furrow was cut off. On the 19th a storm brought temporary relief but after two days without water the horses (magnificent black Australians) and mules had to be turned loose. Only one, a valuable mare called Intombe, belonging to Lieutenant Lindsell, was kept within the ditch. (She survived the war though twice wounded but died afterwards when an attempt was made to remove a bullet.)

But the need of the besieged was desperate. 'Water was everything to us,' noted Winsloe and he offered £25 to whoever dug a successful well. Just before

Christmas, close to the guns in the north-east corner, a decent supply was found. Their most pressing problem was solved.

Meanwhile Boer numbers increased substantially. At one time there were over 2 000 in the town. Cronje could, in Winsloe's estimate, 'have taken the fort over and over again'. Uncomfortable it might have been to try but they could have done it easily. But after the first day no head-on assault was forthcoming. Cronje probably did not want substantial casualties (charging into the face of cannon firing shrapnel was not a pleasant prospect) but he was also reluctant to give up men who were needed urgently elsewhere to thwart relief efforts being mounted by General Sir George Colley from Natal. Because there was something the fort had that Cronje coveted. The nine-pounder guns. He had no artillery. Had he had them the fort would not have held out a day. *Something precious.*

After two weeks things were beginning to look grim for the defenders. There was water and if you kept your head below a parapet you were relatively safe except for a freak intruder bullet or unfair ricochet (if you were careless and raised your head above the skyline a sniper was likely to punish you). But the month's supply of food (tinned beef, hard tack biscuits, flour, tea, coffee, sugar, lime juice for scurvy, and so on) that was supposed to have been kept was not up to the mark and the supplement of erbwurst (a long-lasting German sausage) was unpopular, not least because it was usually accompanied by diarrhoea. There was only a small supply of special foods for the wounded and sick.

The situation was not helped by the use early on of crates and sacks of food to heighten the parapet. The Boers plugged them with so many bullets the iron rations were in danger of producing lead poisoning. A regimen of sandbag-making was introduced using every bit of available cloth but most of the inhabitants, including the women and children, had entered the fort with only the clothes they stood up in. Winsloe got immense satisfaction counting the sandbags as they mounted, likening it to counting game after a shooting party.

Ammunition, too, for both the Martini-Henry rifles and the guns was not as plentiful as could have been wished and had to be expended with circumspection. For the men tobacco ran out very soon.

Christmas, too, was not a happy time. The horse meat that had been set aside for a celebration turned out to be rotten. But the night before, Christmas Eve, a party of four women and seventeen children made an attempt at a night escape, under the guidance of lieutenants Browne and Dalrymple-Hay. But, within a few yards, the smaller children, scared by the silence and the unusual circumstances, began to cry. The Boer sentries, not knowing the composition of the group, opened fire. A babe-in-arms was wounded and young Herbert Talbot Taylor killed. The women decided to press on (and ultimately got clear) but the officers carried the boy's body back to the fort, having promised the distraught mother they would mark his grave for her to find. On the modern monument in the small graveyard that stands near the remains of the fort his name is commemorated but his age is not given. He was just ten.

The first day of the new year brought a shock for the besieged. The Boers had dug up a 100-year-old naval cannon which had been buried at the time of annexation. This was now trundled into service, the *pièce-de-rèsistance* of a new onslaught backed by 1 500 rifles. The venerable old lady, affectionately nicknamed 'Ou Griet', fired round balls of solid lead weighing about five pounds. One of these fell at the feet of Mrs Forssman and is preserved in the Potchefstroom Museum. I have seen several people test its weight in their hands for the first time and all have exclaimed spontaneously in surprise. *Something heavy.*

The Boers had set up their positions in the line of houses closest to the fort (particularly, Jooste's and Dr Poortman's). Now Ou Griet was placed near Steyne's house, protected by sandbags, and she started firing at first light while other attackers quietly infiltrated the cemetery before unleashing a fusillade. Unpleasant as it was to have lumps of lead launched at them, the British responded with successive volleys and after some 20 shots Ou Griet was knocked over by the nine-pounders and it soon became apparent that a charge by the Boers would not be successful.

After this, the strategy of the Boers changed to a slow strangulation by siege. The focus of the fighting shifted towards the cemetery and an unoccupied stone magazine, 200 yards to the north-north-west of the fort. On 3 January 1881 Lieutenant Lindsell occupied this strongpoint with 20 men of the Mounted Infantry. A triangle of action was thus formed – fort, magazine, cemetery. The British dug a sap trench to link their two redoubts while the Boers pushed their saps to within 100 yards of the magazine where they constructed an earthen fort, but their attacks on the tough magazine, though backed by Ou Griet, failed.

Conditions in the fort were unenviable. Three hundred people were confined in a square whose sides were each only two or three yards longer than a cricket pitch! The stench of the ditch was at times unspeakable. Heavy rain (it was one of the wettest summers in the town's history) washed away the earthen walls and compelled constant running repairs. In the dry intervals the sun was merciless.

An hour before dark the women would dine with the officers before retiring to their 'stronghold', a low sandbagged dugout, 9 foot square and 5 foot high, which they had to crawl into through a hole 2 foot by $1\frac{1}{2}$ wide. Here they uncomplainingly passed most of the day and night and did not leave it without permission. The Ulsters and tents they and the wounded used as overhead cover were eventually so riddled with bullets that they resembled the kind of towels provided on Sunday nights in a C-class hotel, bits of tangled string rather than seamless cloth.

The rain and the sun brought another problem for the British. The surrounding grass and the mealies in the fields grew as high as the parapet's eye, providing ideal concealment for the surreptitious Boer movements. The British tried to burn the grass off without success. *Something tall.*

Inevitably, casualties mounted. Corporal Gartshorne, shot through the head, died with the word 'Heaven' as his final statement. Driver Walsh, with a thigh smashed to pieces, only died after ten days of total agony. Private Colvin was hit by an 'explosive bullet', the use of which Winsloe strongly condemned. Even Winsloe's servant, Cruikshank, was hit. Surgeons Wallis and Sketchley amputated his arm (successfully under chloroform), their lamp attracting Boer bullets like a swarm of buzzing insects.

Most grotesque of all was an incident on 15 February when a round five-pound ball penetrated the defensive wall of the magazine and took off the head of Driver Larkins. For a while during the day his comrades had to live with his headless corpse before being able to bury him in darkness. No wonder Lieutenant Rundle asked Winsloe for permission to stand on the walls and 'screech' in order to release his pent-up tension. After warning the sentries, Rundle did just this, sending forth a series of 'most unearthly yells' – something he repeated on several occasions, eliciting withering fire from the Boers.

Inevitably, too, disease, like a thief in the night, slipped past the guards. Lovely, lively Emelia Sketchley, married only a year and pregnant, contracted typhoid fever. Weak from over two months of near starvation she could offer little resistance. She died on 28 February. Surgeon Wallis arranged a truce with Cronje and asked him for a coffin. Next morning one arrived, filled with roots of stephanotis and other flowers. The funeral was held in the open in full view of the Boers, the service conducted by Lieutenant Browne. The watches of the British were not in the best order and when the appointed hour of the truce was up, they were still filling in the grave when a shot from Ou Griet reminded them of the correct time and sent them scuttling for the trenches.

The agony of the Forssmans was not over, however. Two weeks later their nine-year-old son Alric went down with enteric fever. He had to be isolated on a bed of straw outside the walls of the fort but, watched over by his father, he died on 13 March.

Most distasteful of the events of the entire siege were the fates of Van der Linden and Woite outside the fort. They were both accused of spying for the British at the Paardekraal meeting in early December. On the last day of 1880 the former, found guilty of treason by the Boer's War Council, was taken down to the river at the edge of town, made to kneel in front of a freshly dug grave and had a white hood forced over his head. A firing squad of six (three supplied with bullets, three with blanks), positioned closely in front of him, fired three shots into his chest. He was finished off with a fourth bullet to the head by the commandant in charge.

Woite, father of nine, had been betrayed by the carelessness of Clarke. When the special commissioner surrendered the landdrost's office to the Boers they found Woite's letter describing the goings-on at Paardekraal in his pocket. Despite a hurriedly gathered petition of some local citizens and a protest to Cronje from Clarke, he was taken down to the river at 2 o'clock on 6 January. He, too, was shot in the chest and in the head. On the wall of his cell he had managed to record his last thoughts: 'I trust in Jesus. I have given myself into his hands. I am glad to die; in Christ I shall live. In a little while I shall no more be seen; I go to the Father.' He was eventually buried in the garden of his own house.

The British were not entirely passive: on the night of 7 January a small party of six under Lindsell, and covered by 20 more under Dalrymple-Hay, crept up to the wall of the cemetery and poured three volleys into some Boers who were digging entrenchments, sending them running. Lindsell gave ground as a counter-attack came from Jooste's house.

On 22 January a dozen men under Dalrymple-Hay dashed across open veld to attack a trench which threatened the rear of the fort. Three of them were cut down before they had covered a few yards but the rest dislodged 18 Boers, 11 of whom fell as they fled and 4 of whom were taken prisoner (later to be swapped for 4 British prisoners). In a bizarre attempt to trick Winsloe the Boers sent in a counterfeit telegram, purportedly from Bellairs, saying he was coming to the rescue, would arrive on the morrow and expected Winsloe to march out to meet him. Winsloe was not taken in.

For three months, however, from December 1880 to February 1881, the fort was as isolated as any enteric case, a small world wrapped in itself, assailed by an implacable enemy. Except for the curious case of the spy Hudson, almost no news came in. Which was perhaps just as well because none of it was good.

On 20 December 1880 a column of the 94th Regiment, marching from Lydenburg to the relief of Pretoria, was cut to shreds at Bronkhorstspruit, suffering 263 casualties. On 28 January 1881 with General Colley under desperate pressure to relieve Pretoria, as well as other sieges at Rustenburg, Marabastad and Lydenburg, the 58th Regiment was repulsed at Laing's Nek with the loss of 197 men. Finally, on 27 February, Colley himself was killed, together with many of his soldiers, on Majuba hill, and the last hope of any immediate relief was gone.

Though the number of fort-dwellers dwindled through death, release (the Misses Malan and Wart were allowed to leave) and escape (30 African mule and ox drivers voluntarily slipped out on the night of 23 January, though many were shot by the Boers), by mid-March the food shortage was critical. On 20 March Dunne reported that there was left only 1 600 pounds of mealies and 5 000 pounds of millet (hardly fit for animal let alone human consumption), 300 pounds of bran, 24 pounds of preserved meat, 16 pounds of rice and 40 rations of erbwurst, as well as 69 bottles of brandy and 45 bottles of port.

On 6 March General Evelyn Wood and General Piet Joubert agreed to an armistice at O'Neill's cottege near the small town of Mount Prospect in Natal. But on 10 March Ou Griet was set up in the rear (the western side) of the fort in Potchefstroom. This elevated position made things even more uncomfortable for the defenders.

A couple of days afterwards began the most controversial series of events of the whole war. Cronje failed to inform Winsloe of the armistice (he was obliged to do so as one of its terms) and thereby tricked the colonel into capitulation.

On 19 March, knowing that 'the game was up', Winsloe sent a letter to Cronje at sunrise proposing a meeting at noon. At the appointed time a group of mounted Boers emerged, accompanied by a scotch cart. From this they took out a small tent, which they pitched midway between the gaol and the fort, and a hamper. This last item set the British officers salivating with visions of French brandy and other treats, but they dressed as smartly as their circumstances allowed and presented themselves with some dignity before the more dishevelled Boer leaders, one of whom exclaimed, 'How are you all so clean when you come out of that hole?'

In the tent they were duly entertained with choice brandy, biscuits and cigars.

On the following day terms were finally agreed (Winsloe would not give up the civilians with him, some of whom would have been liable to severe punishment, even death). The field guns and rifles were to be surrendered, but the ammunition for the guns would be taken with them by the British.

On the 23rd, after an epic siege of 99 days, the garrison marched out and, at the water-furrow, opened their ranks and laid down their arms before the Boers, 'a fine soldierly lot of men,' thought Winsloe, numbering about 400. After being courteously entertained by the Boers – including dinner with Cronje at the Royal Hotel – Winsloe and his ragged but proud band began their semi-triumphal march through the Orange Free State to Natal. They left behind a restored republic which would retain its independence for 20 years.

There was, however, an extraordinary footnote to the siege. Because of the trickery involved in the 'Treaty of Potchefstroom', the British government refused to accept the validity of the capitulation. They demanded (with success) the return of the nine-pounders and the Martini-Henrys. Unbelievably, too, the British demanded that they be allowed to reoccupy Potchefstroom as a symbolic gesture revoking the capitulation.

So, in late May, General Redvers Buller set out at the head of a column comprising companies of the 94th, two squadrons of the 6th Dragoon Guards and a half squadron of the 15th Hussars, together with a detachment of N/5 Battery (probably under Lieutenant Rundle), and made for Potchefstroom, with sealed orders. They rode, with full military pomp and ceremony, into Potchefstroom on 14 June, passing as they did so many of the shell-damaged town buildings.

When Buller opened his sealed orders he found he was required to occupy the fort, raise the Union Jack and then abandon the fort! Epic was turned into 'comic opera' (Buller's words). It became farce when it was discovered that the British force, not having been forewarned of their mission, had not brought a Union Jack with it. The embarrassment was only relieved when it was remembered that the flag which Clarke had taken down when he surrendered was still in the safe of the landdrost's office. On the 17th Buller marched out again!

Today, the site of the siege is definitely worth a visit – Potchefstroom is an easy drive of less than one-and-a-half hours from Johannesburg. The museum has an excellent collection of documents and illustrations of the town as well as the siege. The staff is exceptionally friendly and helpful and you can hold the round lead ball fired by Ou Griet in your hands.

The gaol, the landdrost's office and the Forssman home have gone, but the house of Landdrost Goetz is preserved as a museum. The old cemetery is still there, minus its walls, but with some interesting gravestones. The magazine that exists today is not the original – it is smaller than its predecessor – but it stands on the same spot and is constructed out of some of the original stones. Perhaps its boundaries are paced by the ghost of Private Larkins forever searching for his head.

Above all, though, the earth banks of the fort still stand in the open. In the little walled graveyard nearby lie Emelia and Alric Forssman and Herbert Taylor and most of the other 25 British dead. Unlike the plaque on the magazine the plaque here has not yet been stolen, though a broken supermarket trolley and assorted burnt debris disfigure the sanctity of the place.

But you can stand on the walls of the fort (where Rundle emitted his terrifying screeches) and work out the geography of the battle. An overpass road, a railway line and a grain silo obscure the view of the cemetery and the magazine but it is easy enough to plot or imagine the magazine, the cemetery, the water-furrow, the goal and even the line of townhouses and the landdrost's office as the were at the time of the siege. The outlines of the gun emplacements on the north-east corner can also be made out.

The most amazing thing, however, is when you look inwards from the walls of the fort. You begin to realise the extraordinary nature of the 99-day siege. The fort is not much larger than a tennis court.

Something very, very small.

CHAPTER TWENTY

— • ◆ • —

BLOUBERG

The *walt hoorns* – war horns – were sounding from the north. The drums of Malaboch were beating. The chief had not come down from his mountain for three years. Now an erroneous charge that he had not paid his taxes since 1881 was the excuse being used by the South African Republic to bring him to heel. It was May 1894.

General Joubert was to be commander-in-chief of the Republic's force. Commandos were ordered up from Middelburg, Rustenburg, Marico, Waterberg and Zoutpansberg. The Pretoria Town Contingent consisted of 175 men, most of whom were Uitlanders. They were joined, from the comfort of Mrs Simpson's boarding house in Proes Street, by the Anglican clergyman Colin Rae. His memoirs of the campaign, evocative and sometimes humorous, were published so soon after the event that it has the smell of fresh bread (copies can sometimes be found in second-hand book stores nowadays but they do not come cheap).

The ragtag force straggled north in desultory and sometimes comic fashion. Rain pelted down, forcing a dozen men to sleep in each wagon. They were more in danger of shooting each other through indiscipline and ineptness. Oxen were lost, so Rae's wagon had to rope in a black cow on one occasion, though even that managed to escape!

Their progress northwards can be followed with some interest by road. Not by the new antiseptic toll road where you pay only for homogeneity and blandness but via the old road which meanders through the small towns, where each place of note was punctuated with an event of some significance.

At Hammanskraal one of the conscripts suffered from drinking bad water and another walked 30 yards into the bush and stabbed himself to death because 'it was hard for him to leave his wife and family and be pressed into service'.

At Pienaar's River they were already missing 18 deserters.

On the approach to Warmbaths, with its hot springs reputedly harbouring medicinal properties, Rae's companion shot a 'steinbok'. He dismounted and put his gun down to go in search of it. Whereupon the horse bolted. Rae followed it on horseback and eventually secured it. While they looked for the buck his companion forgot where he'd left his gun in the long grass. Eventually, after much thrashing about in the undergrowth, horse, buck and gun were all safely united.

At Nijlstroom Rae had some fun at the expense of the Boers and their naive belief that the source of the Nile was located in southern Africa (the unmistakable pyramid hill – known as 'place of spirits' to the Africans – is the one landmark that the toll road passes closely by).

At Naboomfontein an African in charge of a wagon had it run over his leg above the ankle, completely severing sinews and tendons, rendering it useless, presumably for the rest of his life.

At Moorddrift one of the Africans pinned a hare that ran through the camp with a well-aimed chopper and they got fresh water at Biltongfontein.

Finally, after moving through the beautiful scenery of the Waterberg and the countryside beyond, they came within sight of their destination, the mountain stronghold of Malaboch called Blaauwberg or Blouberg. That night Dr Tobias held an al fresco concert where he sang for the first time the 'Malaboch War Song' which he had composed.

> *Ons gaat ten oorlog,*
> *Malaboch! Malaboch!*
> *One gaat jou haal*
> *Je moet op-betaal,*
> *Ons zal jou skiet*
> *Op commando van Oom Piet.*

In the flatland south of the mountain, beyond the mission station of Revd. Sonntag, the Pretoria contingent assembled its laager. The picnic was over as they prepared to 'breakfast off Malaboch'. At nights they were frozen with the cold.

Malaboch's 'hoofstadt' was located high up on the mountain which rose steeply above the plain. He was believed to have between 600 and 1 000 Mananwa warriors with him as well as numerous women and children. The Boers when all their commandos arrived could count on about 2 000 men as well as hundreds of African auxiliaries. While Malaboch hoped for support from Makgato in the Zoutpansberg, he was also opposed and undermined by his enemies Kiviet, Mapen and Mathala on the edges of the Blouberg. The Boers' superiority in firepower was backed by some artillery, including a 7-pounder and a small mountain gun as well as at least one Hotchkiss. They had also brought with them a weapon so monstrous that many people thought it should not be used against humans.

While Colonel Ferreira had tried to instil some discipline in the ranks of the Pretorians, some of the other commandos were motley outfits.

Although the main force had arrived on 12 June after a march of 21 days and were roused at 3 o'clock next morning, the main attack was postponed until the following day, although firing took place throughout the first night.

Joubert (who had suffered much from a poisonous spider bite on the way) decided on a plan of attack from both sides – the Pretorians and Zoutpansbergers were to attack from the south-west and the Waterbergers from the opposite side.

The advance was farcical – without order or method. The majority of the 'feather-bed soldiers' lay down in small parties before they had moved 200 yards. They were, however, met with such little opposition that when the combined forces met on the summit of the designated koppie they placed the mountain gun in position and set about constructing a fort from loose stones that lay plentifully to hand. Forty men were left to defend what turned out to be the key position of the battle that stretched over the weeks ahead.

The eventual strategy that was developed was for a set of forts (perhaps six in all) to be built on the mountain with the aim of squeezing Malaboch slowly like a klipspringer in the coils of a mountain python. Rae thought this was a mistake. He felt that the attitude of the Malabochians, seldom the aggressors in the fighting, was 'nothing more or less than the gentle protest against what they considered an unjust encroachment on their ancestral rights' and he firmly believed that had Malaboch's life been guaranteed from the first then both he and his people would have surrendered without a shot being fired.

Failing this, Rae thought an immediate all-out attack should have been undertaken. He did concede, however, that Joubert's patience and tact might have been necessary to keep together the independent-minded bunch of irregulars he had under his command.

A heavy engagement involving all units was undertaken on 20 June, starting at 2 o'clock in the morning. The Pretorians moved out of their laager and followed a zigzag African path through bush so thick it could have concealed thousands of warriors and in which the Malabochian assegais would have been particularly deadly. Fortunately, no such ambush materialised and they arrived at the base of the mountain immediately below the big spur. They climbed some 700 feet in similar bush before coming to the foot of the koppie they were aiming to capture. Suddenly they were in the open, crammed together and exposed in the bright moonlight. Volleys from the cliffs above poured into them but by a miracle there were no casualties. The koppie was taken and for ten hours fighting ranged over large areas of the mountain. At times the Boer forces found themselves in decidedly threatened situations as new reinforcements arrived on Malaboch's side. Individual escapes were frequently near-things, with several hats and coats acquiring new holes. In another sector the Waterberg commando had advanced with two guns, but the defenders attacked them with such ferocity that one of the cannons was lost and was only recaptured with great difficulty. The Waterbergers had to retreat for fear of being overwhelmed and lost one man killed and six seriously wounded (in addition to ten of their black allies) in the process.

The whole thing had turned into a shambolic defeat.

It was made worse by a reluctance on the part of the Boers (or so it seemed to Rae) to amputate until the last minute. Thus the leg of one Nel, hit in the ankle, turned gangrenous before anything was done. Too late. He was hastily buried, in cotton blankets, under a marula tree in the camp. The night before, it being particularly dark, one of the medical staff, Mr Schmidt, had fallen into the open grave, so seriously damaging his knee-cap that he was unable to walk for several weeks.

With the enemy occupying most of the koppies the besiegers were not very happy. Provisions and clothing were their major grievance. Sugar had run out in the first week of the march and the lack of vegetables meant a number of cases of veld sores. Many of the men were literally in rags. Piet Zeederberg, designated scavenger, left camp to scour the Zoutpansberg and could purchase no more than 100 bags of meal in the whole district.

Fighting on the mountain was desultory. Malaboch's defences at the top of the mountain were strong, situated as the stad was on a precipitous kloof protected by immense boulders and thick bush, and made almost impassable by well-made and concealed schanzes. Revd. Sonntag attempted to see Malaboch. A truce was arranged but was broken by some young Boers on outpost duty who were not aware of the ceasefire.

The distress of the defenders was awful. Food had all but disappeared: water was extremely scarce. The aged, the women and the children suffered most.

The Pretoria fort, high up on Blouberg, was about seven-and-a-half miles from the laager 2 700 feet down below. The countryside in between was extremely beautiful but equally dangerous. There was no protection along this route for those conveying supplies other than the 25 men at Fort Jonker, which was 1 500 hundred yards off the path, and an outpost of some 30 African friendlies two miles further on.

So it was with great trepidation that Rae found himself at 4.30 in the cold morning of 7 July accompanying a procession 100 yards long in which 100 bearers were transporting 53 boxes up to the fort. Each box contained 50 pounds of the much-deprecated (though in earlier wars, enormously effective) weapon – dynamite (patented in 1867 by the Swede Alfred Nobel and readily available in a gold mining country). The steep six hour climb was difficult for each of the bearers with 50 pounds deadweight on his head. And one well-aimed shot from the dense bush might detonate the lot!

Six men under Lieutenant Schroeder were allocated for the job of mining on a spot directly above the most important caves of Malaboch's *hoofstadt*. A charge of 350 pounds of dynamite was laid and a bugle sounded as warning. The occupants of the nearby fort scattered for cover as Schroeder lit the three-and-a-half minute fuse. The ensuing wait seemed like eternity.

Then a huge cloud of smoke and dust, accompanied by a bang as if 'a dozen 100 ton guns were fired simultaneously' made its presence known. Huge fragments of rock shot into the air, so high they were almost lost to sight, and made a frightful whistling sound as they announced their return.

Scouts went to assess the damage. Though a large section of the krantz had shifted and a lot of huts had been crushed under the debris, the main cave itself seemed to have remained unscathed.

Subsequent efforts to blow it up proved equally fruitless. So much for the secret weapon! Indeed one of the dynamitards, Joe Morris, scored an own goal when a piece of falling rock hit him on the head, causing a scalp wound. What did terrify the Malabochians were the rockets.

The defenders were not entirely quiescent. It became clear that several traders were cashing in by supplying them with ammunition and superior weapons. The earlier muzzle-loaders (and even stone bullets) began to give way to Sniders and the easily recognisable dull boom of the elephant guns gave way to the menacing sharp ping of the Winchester express rifles.

On one occasion the defenders turned the tables on their persecutors. A detachment of Pretorians had occupied a small fort on a ledge some 300 yards lower down from Fort Pretoria. Most of the men sheltered themselves from the bitter cold in cosier nooks amongst the boulders a short distance from the fort. They had just settled themselves inside the picket lines at about half past seven and all was still when two volleys of bullets suddenly poured in upon them. Surreptitiously the enemy had crept in crescent formation to within 20 or 30 yards of the pickets. Amazingly only two of the Boers were lightly injured, and they all scrambled for cover in the fort where they put up a regular fusillade of bullets and threw hand grenades into the surrounding crevices. No fire was returned; the intruders had melted away but the rest of the night was spent in sleepless anxiety.

On 27 July the Malabochians created a diversion by attacking the artillery laager (at the base of the mountain). The 'friendlies' in the laager panicked and left it in a hurry, only 12 of their number remaining behind to help the artillerymen who had trouble dragging their gun around to train it on their attackers, who kept up a steady fire. After two hours of warm fire the attackers withdrew and the Boers were lucky to have escaped without casualties.

But casualties were mounting. When a party which was constructing a fort had to retire when it came under fire the rearguard of six men found themselves in a perilous position. Schmidt, a relative of the man who had portentously fallen into Nel's grave, was terribly wounded in the loin; Dunsdon was struck on one side by a bullet and on the other by a charge of buckshot; Cowley had a ball go straight through his stomach and out the other side; Scobel's rifle jammed and Keith had only two rounds left. Fortunately their companions fought their way back to relieve them. But Schmidt died. He was born in Baden, Germany, and was 19 years old. He was luckier in death than Nel since by this time coffins made from packing cases were available.

Groenewald's death was grotesque. He noticed a beehive on a ledge above him, and raised himself up, determined to take it back with him. He was immediately shot through the heart and his body fell back into some huts that the Boers were burning in the *hoofstadt*. The cartridges in his bandolier began to explode like a string of firework crackers and all his companions found of him were his charred remains.

Lottering's death was slower though no less terrible. A bullet had entered at the elbow and passed all the way up the arm, smashing the bone to splinters before finally passing out at the shoulder. For a long time the doctor, without a sufficiently long probe, could not find the passage of the bullet. After three hours he was about to give up when his assistant offered him a ramrod which did the trick. Tubing was then attached to the top of the rod and carefully drawn through, so that a constant current of filtered water was kept running through the limb.

However, days later, conscious to the end, though unable to speak, Lottering died. The artilleryman was buried in the artillery camp.

Both chloroform and morphine were available and eased some suffering but survival chances were not helped by the Boer procrastination over amputation.

Eventually the Boer tactics of strangling Malaboch and his followers with a series of forts and denying them access to water began to pay off. One day, towards the end of July, over 100 women and children emerged from the caves around the *hoofstadt* and gave themselves up. Many had to crawl to water containers and had to be forcibly restrained from drinking themselves to death.

At 5 o'clock that evening the Boers moved to form a semi-circular cordon around the main cave, keeping some 70 yards from the entrance. The goal was to pen Malaboch and his people inside. At 9.30 the besieged sallied out to find water but were driven back with a furious fusillade of some 600 rifles firing simultaneously.

Then, clear in the dark night, a voice rang out, distinctive and musical. The air of authority left no one in any doubt as to the speaker and the sharp words reached the forts all of 250 yards away.

'You have taken from me my women and children, my cattle and corn,' said the chief. 'My villages have you burnt and now you will not even let me have a drink of water; everything that was mine you have, wait until tomorrow and you shall have me. What do you seek in fighting tonight?'

To this the interpreter replied, 'The wolves and jackals come out from their holes only when darkness has covered the land, so they must be hunted during the night. Come forth in the light of day, and we will hear what you have to say; but if you will not face the sun, then we must shoot you as we shoot the jackals.'

Malaboch's reply was awaited, but none came, except a request to see Revd. Sonntag.

Next morning the missionary engaged in a long conference inside the cave. Then, after a shouted plea to hold fire, the women and children and aged, 493 in all, hobbled out or were carried out. The horrors of war were all too visible in the emaciated faces, thin bodies and rotting limbs. The stench of scores of dead bodies wafted out from within. Tales were told of babies killed by the shock of exploding shells.

But Malaboch did not come. Only on 31 July did he surrender (accompanied by four grizzled indunas and two young boys who were his sons) and was taken to the Rustenburg camp. He was about 30 years old, five foot ten in height, with a small moustache and beard. He wore a light corduroy suit but no hat or boots.

Rae was suspicious. When the cave was approached to flush out the remaining warriors none was found. They had all escaped into the bush. Was then the man who had surrendered really Malaboch?

That night the captive chief seems to have tried to kill himself by flinging himself face first into a fire.

The position of the two clergymen makes an interesting contrast. While Sonntag was of the opinion that Malaboch had been 'most unjustly treated', Rae, a war cleric, was more convinced of the righteousness of the cause and, being less a feudal crusader than a representative of modern industrial progess, felt the chief's power must be broken and laziness replaced by hard work. But the chief's fate was not in their hands. A Council of War deliberated his fate. Some felt he should be executed forthwith, others that he had surrendered under promise of his life. Eventually it was decided he would be kept until the end of the campaign and transported to Pretoria.

A carved 'god' in the shape of a crocodile as well as war drums and other weapons, marked with the sign of the crocodile, were also brought into camp, as the spoils of war.

Rae and his compatriots turned for home. The highlight of their trip was croquet and a concert at the Warmbaths Hotel where Dr Tobias sang all eight verses of the Malaboch War Song:

> *Our Commandant General –*
> *Malaboch! Malaboch!*
> *He could shoot you at sight*
> *But prefers dynamite!*
> *You may well go in dread*
> *With the stones thrown at your head.*
> *(Chorus)*

After 12 months' absence the Pretorians were officially welcomed by all the town and government dignitaries and unofficially by a great throng of vehicles, cyclists, horsemen and pedestrians. President Kruger was particularly complimentary to the Uitlanders: 'You have willingly gone, and willingly entered into the rocks, into the dark caves, and brought forth the enemy … By doing this you have given proof of your faithfulness to the South African Republic, the State you have accepted as your fatherland.' He promised to draw up a list to submit to the Volksraad of those who might be considered for the franchise.

A few months later the Jameson Raid changed the political climate.

CHAPTER TWENTY-ONE

MAGOEBASKLOOF

H ere's a historical mystery tour for you. And it involves no real hardship since it entails visiting one of the most exquisite and tranquil parts of South Africa.

When the Portuguese of the fifteenth century set out to discover a sea route to the east round the south of Africa, wealth was undoubtedly their main aim. But they had other goals, too: one was to outflank the Muslim world and link up with Prester John, a supposedly Christian king of Ethiopia (Abyssinia). I wanted to pursue Prester John.

So it was to Magoebaskloof, east of Pietersburg in the province of Limpopo, that I recently took myself. But Magoebaskloof is a long way south of Abyssinia, you might exclaim and, shaking your head in dismay and disbelief, move off to a livelier and better-informed travel guide.

For those few readers who plod through everything without discrimination and who consequently remain with me, I shall try to establish a connection and may even persuade one or two of you to pursue the fascinating journey I undertook.

The gateway to Magoebaskloof is the village of Haenertsburg. The road there from Pietersburg is as deadly as anything in Wole Soyinka's play *The Road*, especially when it is presided over by a mad professor. The countryside in between is harsh and rocky and dry, the poverty dire. But the entrepreneurial spirit does occasionally manage a xerophytic existence, even on stony ground.

Beyond a place called Nobody a sign with the buzzword 'tourism' caught my eye and attracted me to the shack that Philemon Mmini has constructed out of cans and empty bottles. Equally conspicuous alongside it is his pyramid-shaped p.k. with the sign 'Long Drop Toilet of Cans' painted along the side.

Mr Mmini deftly makes model motor cars out of scraps of metal. Pretty crude stuff it must be admitted but, as he says, 'hunger helps you think of things to do and ways to make a living. Living next to the highway I thought I could sell things to tourists. This area here is too poor and the guys do nothing'. He came to this arid place from Springs after being fired for reversing his dump truck into the boss's car. Now from the meagre spoils of tourism he employs one apprentice and pays R80 a year to the local chief as rent for the plot.

By the time I got into the vicinity of Haenertsburg the landscape had begun to change to lush woodlands with waterfalls and hikes and nurseries and friendly pubs, a land of silver mists and mysteries and mellow fruitfulness.

Just beyond Haenertsburg is the real object of our interest. The road straight on leads to the wonderful valley of Magoebaskloof. A road leading off to the right descends into the parallel valley of the Great Letaba (or George's Valley) to the south. A huge ridge which used to be called Letaba Ridge separates the two valleys. A few kilometres along the George's Valley road (the S528), still on the escarpment, is a great rock dedicated to the memory of John Buchan, famous Scottish writer of *The Thirty-Nine Steps* and at one time Governor-General of Canada.

Buchan came to South Africa as a young man in 1901, working for Lord Milner, and stayed for two years. One of his principal jobs was to resettle Boer families on the land after the war. He managed to make several longish journeys around the Transvaal and Swaziland but the one that made the most impression on him was to the north-east, to the Wood Bush and Magoebaskloof.

Under the Wolkberge and Iron Crown mountain he rode along the escarpment with its open, rolling hills and valleys of dense indigenous bush, reminding him somehow of his native land. 'I have never been in such an earthly paradise,' he wrote. 'The whole place looks like a colossal nobleman's park laid out by some famous landscape gardener. And when you examine it closely you find it richer than anything you can imagine. The woods are virgin forest ... The valleys have full clear streams flowing down them and water-meadows ... The perfume of the place is beyond description. The soil is very rich: the climate misty and invigorating, just like Scotland.'

He was fascinated by, fell in love with the area. It was a sacred grove of the gods, 'a temenos, a place enchanted and consecrate'. He was, he said, 'very keen to have a bit biggin of my ain in it' and determined one day to build a lodge there in the wilderness. But he never did.

He did, however, eulogise the place in his first book *The African Colony* published in 1903 and he used it as the setting for the country house in his anonymously published *A Lodge in the Wilderness*, a curious book depicting a retreat-like symposium at which thinly disguised, well-known statesmen and aristocrats (including Milner) discuss Imperialism. The prominence he gave the area in his autobiography, *Memory Hold-the-Door*, published in the year of his death, also indicates how formative and haunting was the experience.

The memorial stands at the head of a magnificent prospect, looking down into the great valley of the Letaba River and overlooking the modern Ebenezer Dam. A plaque carries part of the appropriate, intensely beautiful passage from his autobiography. I shall not spoil his words for you by repeating them here. There is no substitute for reading them *in situ*. That experience alone – combined with the view – makes the visit to the whole area worthwhile.

It is not all, of course. 'When you get to the edge of my water-meadows,' he also wrote, 'you look down 4 000 feet to the shimmering blue plains of the fever country ... One day I rode down into the fever country – appallingly hot but very interesting. I slept New Year's Night there in the house of a German called Altenroxel where I ate the most wonderful tropical fruits and saw a lot of snakes.' He also witnessed down there some kind of ceremony, probably a circumcision school, which was 'enormously impressive' and which may well have been part of the inspiration behind his most famous work set in the area – his first novel, *Prester John* (1910).

A very brief summary of the book is necessary. David Crawfurd, a young Scottish boy, surprises a visiting black clergyman performing some atavistic rite on a dark beach in Scotland. When Crawfurd comes to the Transvaal to run a trading store somewhere between the Wolkberge and the Zoutpansberg he re-encounters John Laputa now trying to organise, under the guise of religion, all the disaffected African tribes into one revolutionary movement. The secrets of Laputa's power are his claim to be the direct successor of Prester (Priest) John and his access to the fabled king's fetich collar of rubies. The novel is full of drums and rites, caves and conspiracies, passwords and spies.

Crawfurd's journeys take him into the lowveld where most of the blacks live and to a hidden cave and lake in the 'Rooi Rand' (somewhat related to the Zoutpansberg) which is the source of a great river. Laputa's plan is to bring the necklace from there to a meeting of all his followers gathering at a place called 'Inanda's Kraal' in the Wolkberge where he would proclaim his sacred kingship. But Crawfurd manages to steal it and hide it in a pool in a deep valley leading up to the top of the escarpment where most of the whites live. Ultimately he thwarts the revolution, Laputa is killed, a modern school is built and a statue of Laputa is erected in front of it. Crawfurd returns home, à la Rider Haggard, with much treasure he has found in the cave. This is, after all, an Imperial novel.

Nevertheless, it is interesting all the same and the devil is in the details. Buchan's fears were a combination of factors which must here be simplified to two: firstly, the area had been the scene not long before of strong African resistance to white encroachment; and, secondly, in the decade before Buchan's arrival, various religious movements, loosely called Ethiopian churches, had spread amongst Africans. An eccentric white, Joseph Booth, had, for instance, not so long before wandered through the land preaching 'Africa for the Africans'. These religious movements divided roughly into two kinds: those who broke from the established churches, but, somewhat conservatively, retained most of their forms (the Ethiopians) and those who sprang more from traditional practices such as healing or prophecy (the Zionists). Some knowledge of this is essential to the understanding of *Prester John*. Buchan's underlying fear was of a unification of armed resistance and militant Africanist religion.

At the beginning of the novel there is a rough map rather like a treasure map. The challenge of this historical-cum-literary tour is to determine from this and many other details in the book what was fictional and what was factual in the map that Buchan had in his mind. To pursue this you may resort to several techniques of research: you will need modern maps as detailed as possible to make comparisons; you might read books for background (such as Bengt Sundkler's *Bantu Prophets in South Africa* or James Campbell's *Songs of Zion* on Ethiopianism; Jensen Krige's *Realm of the Rain Queen* or Louis Changuion's booklet on the centenary of Haenertsburg for localised material, though the last may be difficult to find; or the books of Harry Klein and T.V. Bulpin); and you should talk with the locals who are extremely open and friendly and only too happy to share their beautiful part of the world with you. Historian Louis Changuion probably knows more than anyone.

I shall not spoil the joy of discovery for you with too much information but one or two clues might help as a guide to your understanding of *Prester John*.

Buchan's Wesselsburg is clearly Tzaneen, so from the Buchan memorial you should drive down the road alongside the great Letaba River, which plays a large part in the novel, to the bustling lowveld town, and especially pay a visit to the town museum. Do not be fooled by its shabbiness: poorly housed, run on a shoestring, deliberately without information labels as it is. Get talking to its curator, Jurgen Witt, and you will soon be caught up in his deep affection for and expertise on the Tsonga people and their art, the Rain Queen, Mujaji (the model for Rider Haggard's mysterious 'She'), and her history. It is always a source of wonder to me how funding bodies, including government, fail to spot genuine gems in the sifting pans of gravel under their very noses.

In the museum, too, you will find a map of the whole area drawn by the Swiss missionary Henri Berthoud in 1903, exactly at the time Buchan was there. This can be most useful in the search for *Prester John* clues.

From Tzaneen take the other road (the R71) back to Haenertsburg. This road travels through Magoebaskloof which Buchan corrupted to Machudi's Glen. Halfway up, the Debegeni Falls are worth a visit, with its shady rock pools in one of which David Crawfurd hid the ruby collar of Prester John.

At the top, from the gardens of the Magoebaskloof Hotel, preferably with a beer in hand, the view down the kloof, often lazy and hazy, is one of the most peaceful and beautiful I know in the whole world.

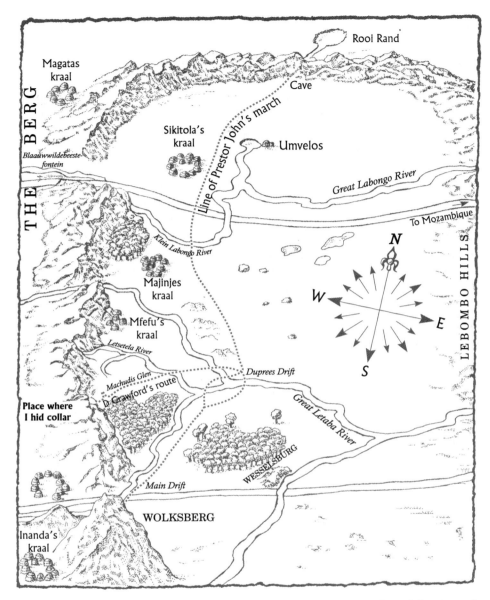

Magoebaskloof (based on John Buchan's original sketch representation of the events in Prester John*)*

It wasn't always peaceful.

As whites began to filter into the area in the 1870s and 1880s and to lay out farms conflict with the local people was inevitable, especially when government policy was to confine the locals into limited 'locations'.

Things came to a head in 1888 when beacons were put up to demarcate farms in the territory of Chief Makgoba. As soon as they went up Makgoba's people tore them down. The chief himself was fined and, when he refused to pay, was imprisoned in Fort Klipdam near Pietersburg. He dug his way out and escaped, and swore never to submit to such an experience again.

Mujaji, too, was unhappy with the location policy and, in 1890, a strong government commando was sent to pacify her but no decisive engagement was effected, the fabled Rain Queen herself evaded capture and rain, incessant rain, appropriately enough impeded the success of the operation. The district simmered.

In 1892 and 1893 there were sporadic attacks on white farms, to the point where the government contemplated a complete abandonment of the lowveld. Instead, however, it warned Makgoba and others to return to their allotted territories. But for the moment they could do little except gather the scattered farm families into Haenertsburg and New Agatha because most of the commandos were engaged in the war against Malaboch and the investment of Blouberg. The tribes of the east were well aware of what was going on in the west and it was feared by the government they would take advantage of its distraction.

Only at the beginning of August 1894, after the defeat of Malaboch, could attention be concentrated on the east. On 12 August, though, a small force under Field Cornet Hendrick Alberts came close to disaster in an encounter with the bands of Mashuti, Mmamathola and Mogoboya and had to take refuge in the village of New Agatha, which had to be defended against incessant attack all night.

When the main commando forces arrived local villages were systematically mopped up, hampered as they were by the rain. In early September they assaulted Makgoba's village. After initial fierce resistance the chief and his warriors backed off and disappeared into the dense indigenous bush that covered the deep kloofs in the area.

The commando then turned on Mujaji and surrounded her home. On 20 September she agreed to give herself up. It was a significant moment; the world held its breath.

'It would be the first time,' writes historian Louis Changuion 'that white people would see the Rain Queen.' What happened was not what they expected. 'After four days,' Changuion continues, 'an old wrinkled black woman was carried out on a litter, accompanied by her chief indunas, to negotiate with the white people. It was a great disappointment to the men watching the proceedings – of "She-who-must-be-obeyed" there was no trace. She was not the white woman of the legends. It is told that [Commandant General] Joubert presented her with a "kappie" (bonnet) and a blanket.'

The wily Makgoba was still free, however, when the commandos disbanded. The strategy was to build a set of forts, the first two of which were at Modderspruit (near Mujaji) and at New Agatha, followed by three more, one of which was at Joubertskroon (where the Magoebaskloof Hotel now is). They did not stop Makgoba whose men continued to raid farms, including that of Altenroxel in February 1895.

So a new commando was called up and assembled in early June at Joubertskroon. It consisted of about 1 000 Boers from Pretoria, Rustenburg, Waterberg, Middelburg, Lydenburg, and Zoutspanberg with some 3 000 African allies. They were to sweep round the valleys in three divisions. To mark them out as friendlies the black auxiliaries were instructed to wear white headbands.

The first attack, on 3 June, broke down. Makgoba's men did not play fair – some of them put on white headbands and nearly trapped their attackers in unsportsmanlike ambushes, using false colours.

Next day a couple of cumbersome artillery pieces were hauled down into the kloof and there was a lot of thrashing about in the dark forest with its narrow paths and dappled sunlight and shade. The commando found and fired Makgoba's hideout but they did not find the chief himself. On 8 June the government forces came at him from four directions, but it was like trying to hit a golfball with a straight stick.

Finally, on 9 June, Abel Erasmus, the Native Commissioner, brought up a body of fearsome Swazi warriors who early that morning swept the forest, flushed out one of the wives of Makgoba and persuaded her to show where her husband had taken refuge – which turned out to be to the east of the Letaba Ridge (which is on your right as you look down the kloof from the hotel) in a remote glen in George's Valley.

He seems to have defended himself bravely, blocking a narrow path before being overpowered, executed and having his head cut off. In *Prester John* his skull makes a brief appearance on the farm of Altenroxel as a drinking cup.

To get the best view of the Letaba (George's) Valley, I visited Patrick McGaffin on his farm Crown Mount. Patrick ('very Gothic' was a neighbour's description of him) is passionate about his habitat, produces a fine crop of green tea, and regaled me with many stories and much information. He believes he has found Makgoba's grave in an almost inaccessible spot close to his farm.

Patrick also showed me something else on the farm. *Prester John* makes a big thing of the road between Haenertsburg and 'Wesselsburg'. The government forces try to prevent Laputa, north of the road, from linking with his followers at Inanda's Kraal, south of the road. Patrols use the road as the equivalent of a fence.

Neither of the present main roads between Haenertsburg and Tzaneen were in existence in 1903. There was, in fact, only the old coach road, remnants of which Patrick showed me on the farm. I saw for myself how perilous the steep descent must have been – as terrifying for the passengers as anything out of Washington Irving and Sleepy Hollow. (In fact, when descending, they took the wheels off and used the coach like a sled.)

Near to the Magoebaskloof Hotel is another must-see. In 1999 a fine monument was erected to Makgoba, the 'lion of the Wood Bush', and within it was embedded a plaque and a bust which was based on a photograph of his severed head. The plaque is still there but the bronze head, no doubt in the interests of economic meltdown, like so many other bronze monuments round the country, was within a short time wrenched from its bolts and disappeared. Fortunately, the bust was recovered before it was reduced to a featureless lump, and now stands in the safer environs of the hotel's grounds, one hopes permanently. To lose one's head, Mr Makgoba, may be regarded as a misfortune; to lose two looks like carelessness; if it were to occur a third time one might suspect you of wild and wastrel habits.

If by now the reader has not peeled off to the region of the Groot Spelonken and the Zoutpansberg to look for caves and lakes or gone to Mujaji's to wait for rain, you will notice that we have completed a circle rather like the loop of a lasso and come back to Haenertsburg where the new war memorial to all local people killed in all our wars is well worth paying your respects to. It is called the Long Tom Memorial, the ceramic insets are simple and effective, and the whole is unpretentious and moving, a fitting end to our tour.

Buchan's novel is full of the – often disturbing – prejudices of his time but there are also plenty of ironies associated with it – ironies which may have become sharper to you as you return along the Pietersburg road – the rope end of the lasso – past the huge Zionist city of Moria on your left, the University of the North on your right and Philemon Mmini's tourist attraction and long drop toilet of cans somewhere in between.

But why all the fuss about Magoebaskloof? Well, that's where Buchan's novelistic career really began and where, in all its mystery and intrigue, his first novel was set.

And why all the fuss about Buchan? Because, with all the spyings and conspiracies – he was after all close to Milner and Military Intelligence – Buchan in his later Richard Hannay novels became the father of the modern espionage thriller and the antecedent of John le Carré, Len Deighton and many, many others. I spy, with my little eye, something beginning with M.

CHAPTER TWENTY-TWO

THABA-BOSIU

This is the story of a mountain. It witnessed many savage battles.

Having been attacked in his mountain stronghold at Botha-Bothe by the Amangwane in 1822 and besieged for three months by the Batlokoa under Sekonyela in 1824, Moshoeshoe began to think of moving his people, the Bamokoteli, out of the path of marauding parties from the north and east to a more remote refuge.

He heard from people related to his people that there was a mountain, some 80 kilometres to the south-west, which might be more impregnable than his present fortress. He sent his brother Mohale to spy out the land and Mohale returned with a most favourable report.

So, in June 1824, in the bitter cold of winter, the Bamokoteli, numbering perhaps a couple of thousand, moved southwards, hugging the fringes of the Maloti in order to escape the attentions of the Batlokoa. As they did so they were harried by a band of cannibals under their chief Rakotsoane, who picked off the stragglers as they fell behind. Among the victims was Moshoeshoe's own grandfather, Peete (the Stammerer).

On the second day they traversed the Khamolane plateau and looked down on the Phuthiatsana valley with the curiously shaped hill Qiloane in the centre and the great plateau of Berea to the west. They also caught their first sight of the mountain which was to become like a mother to them. It was getting dark by the time they climbed the Khubelu (Red) Pass to the top and the mountain seemed to grow larger as they did so. So it was called Thaba-Bosiu (the Mountain by Night). The sides of Thaba-Bosiu plateau or mesa rise 350 to 400 feet above the valley which surrounds it. Its shape is somewhat irregular, approximately two miles long and nearly a mile wide and the area on top is some two square miles in extent. A ring of cliffs around the summit makes access other than via the six passes (all readily defensible) almost impossible and several fresh springs provide ample water for defenders and herds. (I am told by a correspondent – Quentin Coaker – that several sand dunes on the top of the mountain, known as the 'shifting sands' because they were made of the finest sand, were believed, perhaps falsely, to contain the water necessary to sustain the defenders during a seige.) Below it in the valley runs the Phuthiatsana River. Moshoeshoe established himself on top, close to the Khubelu Pass, and his father Mokhachane settled on the south-west corner of the mountain. Moshoeshoe's brother Job built his village in the valley below Mokhachane. Moshoeshoe slowly gathered into his fold refugees scattered by the terrible wars which raged all around – an inclusive system which became the Basotho nation.

The first test of Thaba-Bosiu's defences came under an attack by Matiwane's Zulu-speaking Amangwane (either in June 1827 or February 1828 – the sources consulted differ). The main force of the Amangwane, swarming over the Berea plateau, headed for the Khubelu Pass and the western side of Thaba-Bosiu, while a smaller regiment swung round to the south to attack Moshoeshoe's brother Mohale at Qhobosheaneng.

Moshoeshoe decided to confront the Amangwane in the open (probably to avoid an agonising siege) with the option of retreating up the mountain in the event of things becoming too hot down below. His main army he drew up at the foot of the Khubelu (probably on the ridge where the Lesotho Evangelical Church now stands) while his brothers Makhabane and Posholi did the same at Mokhachane's Pass. His sons, Letsie, 17 years old, and Molapo, 14, were put in charge of defending the passes with an array of rocks and boulders to be precipitated on any luckless foe who threatened the approaches. To delay his enemies Moshoeshoe ostentatiously doctored the Phuthiatsana River crossing places, forcing them to take a longer way round.

The Amangwane, however, were supremely confident – young maidens bearing beer and food watched the action from the heights of Berea plateau, egging their heroes on. As the armies drew up facing one another they exchanged insults as if in front of the walls of Troy before engaging in combat. Although Moshoeshoe's own age regiment, the Matlama, was to absorb the main thrust, he concealed part of it and held it back. The Mollo (fire) section – the first engaged – had to give way until the Liqela ('the beggars') joined them and put the Amangwane to flight.

In the fiercest of the fighting Moshoeshoe's great friend, Makoanyane, like Hector of old, slew at least ten of the enemy in single-handed combat. As they pursued the humiliated Amangwane the best prize was the capture of the provisions their enemy had earmarked for their own celebrations but which their women had forsaken in their own precipitate flight, not wanting themselves to be part of the reverse festivities.

In 1830, the flintlock muskets of the Korana were out-matched by the stone missiles of the Basotho and the intruders repulsed from the heights of Thaba-Bosiu. In 1831 regiments of the formidable Ndebele of Mzilikazi were reported to be approaching and many of the Basotho prepared to stand, not in the open which would have been suicidal against the highly trained warriors, but behind the defensive lines of the mountain itself.

The Ndebele arrived towards dusk and drew themselves up on the far side of the Phuthiatsana. Next day, they divided their force to threaten five of the six passes. Moshoeshoe accordingly dispersed his own army to the danger points, using women and children as beasts of burden keeping the men armed with a regular supply of stones. The hail of stones and avalanche of rocks was so ferocious that even the single-minded Ndebele were driven back. As they retreated, the astute Moshoeshoe sent a herd of cattle after them to feed them on their way back to Mzilikazi and no doubt to appease them sufficiently so that they would not return.

The fearsome Zulu also swept through the area but never attacked Thaba-Bosiu directly and Moshoeshoe was careful to send regular tribute to their king.

It should not be thought that the Basotho were passive only. They themselves mounted significant aggressive raids against the Thembu, for instance, and found their northern neighbours, the Batlokoa under Sekonyela, an irritant until in 1853 Moshoeshoe decided to dispose of the threat once and for all. In a surprise raid he surrounded Sekonyela's twin fortresses of Marabeng and Joalaboholo (magnificently visible as you approach Ficksburg) and broke the Batlokoa power forever (a remnant of the tribe was eventually allowed to settle in the inhospitable interior of the Maloti).

In the 1830s two developments helped strengthen Moshoeshoe and allowed him slowly to integrate the disparate groupings (attracted by its fame to the safety that Thaba-Bosiu seemed to offer) into a unified nation. In 1833 three French missionaries of the Paris Evangelical Missionary Society (PEMS) – Arbousset, Casalis and Gosselin – arrived and their advice (often acting almost as foreign ministers) enabled Moshoeshoe to behave with insight when confronted by the new encroachment by whites. In 1836 Casalis established the mission station at the foot of Khubelu Pass which was to witness many significant events in the course of the rest of the century.

Secondly, with the Korana came the horse and in 1833 a German visitor called Seidenstecher, in return for the hospitality he received, left as a present two horses, a mare and a foal. Eventually that hardy breed – the Basotho pony – was to become as familiar a part of the landscape as the Basotho blanket. Moshoeshoe himself learnt to ride, at first using two long sticks on either side for balance. When the French missionaries first arrived they were greeted by two near-naked men riding bareback into their camp causing a mini-crisis. These apprentices of a new art were Letsie and Molapo, oldest sons of the king.

As they acquired guns in the decades that followed the Basotho developed their skills as mounted infantry perfectly adapted to the broken terrain of their homeland and presented themselves as formidable opponents to any invaders. This General Sir George Cathcart (governor of the Cape) certainly discovered in 1852.

Cathcart had decided to assert his authority in the Orange River Sovereignty by a show of force against Moshoeshoe. Backed by a force of 2 000 men, including cavalry and artillery with six-pounder guns and twelve-pounder howitzers, Cathcart presented the Basotho king with an ultimatum demanding 10 000 head of cattle and 1 000 horses within three days. On 18 December Moshoeshoe delivered 3 500 cattle with a promise of more. This did not placate Cathcart who pressed forward, with the object of seizing the Basotho herds and then arriving at Thaba-Bosiu and dictating terms. He never reached it.

At dawn on 20 December Cathcart's invading army crossed the Caledon (Mohokare) River some 20 kilometres north-west of Thaba-Bosiu. There was a major obstacle in his way – the imposing bulk of Berea plateau. Cathcart's plan was to send Napier to the north and round the plateau with a combined force, numbering some 230 men, of Lancers and Cape Mounted Rifles; Eyre passed the mission station at Berea and over the plateau itself with about 500 infantrymen; while Cathcart himself swept south of the plateau with 300 infantry, some cavalry and the howitzers. He did not expect significant opposition – each man carried only six rounds of ammunition. His intelligence, however, was faulty – the plateau extended much further than was thought and Napier had to make his way to the top, where he was intercepted by Molapo and 700 mounted Basotho who inflicted serious casualties on the Lancers and stopped Napier's advance in its tracks.

Eyre was more successful – he captured cattle but was involved in a nasty incident where some women and girls were killed, whether by accident or design. A few of his men were led into an ambush by some Basotho wearing headgear and lances plundered from Napier's dead cavalrymen. But the disciplined fire of his infantry kept the Basotho at bay and a heavy thunderstorm intervened, allowing Eyre, late in the afternoon, to attempt a link-up with Cathcart, according to his instructions.

Cathcart, meanwhile, had advanced round the plateau, pushing the Basotho before him with his small body of cavalry and his artillery. He halted at a knoll overlooking the Phuthiatsana about three miles from Thaba-Bosiu, nervously waiting for Napier and Eyre to join him. Increasingly, the Basotho strengthened their force in front of him. The thunderstorm temporarily put an end to the desultory skirmishing but when it cleared Cathcart could see how precarious his position was – Basotho mounted infantry numbering between five and six thousand were sweeping around to threaten his right flank. Fortunately for the British, Eyre arrived at this moment and strengthened this flank.

The Basotho, with various of Moshoeshoe's sons (including Masupha) prominent, made several determined charges and were only with difficulty held off. Cathcart retired to the abandoned village of Boqate nearby and, short of ammunition, spent a sleepless night there. Moshoeshoe, too, was worried, knowing that, in the long run, the British were superior in resources and firepower.

That night he and 'his' missionary, Eugène Casalis, devised their famously brilliant letter, addressed to Cathcart, suggesting that he had 'chastised' the Basotho and entreating peace. This offered Cathcart, who had lost 38 killed, a dignified way out and he duly backed off.

Victory was really Moshoeshoe's but, partly as a result of the battle of Berea, the British finally abandoned the Orange River Sovereignty and opened the door to an altogether more implacable enemy of the Basotho.

As the Boers began occupying farms on the west bank of the Caledon which Moshoeshoe, with considerable justification, claimed as his territory, friction increased and open conflict was almost inevitable. It came in 1858.

The Boers set fire to the PEMS station at Beersheba (near Smithfield) and went on to destroy most of Morija, except the church. At Morija they burnt the house of Arbousset, whom they particularly disliked, in the process sending 25 years of research into Basotho history, life and customs up in flames – the worst thing that can happen to a scholar. The ferocity of their onslaught may have been exacerbated by the check they received at the hands of Letsie at Hellspoort and the discovery that a number of their fallen comrades had been mutilated. They arrived in the vicinity of Thaba-Bosiu in May but their meagre strength was confronted by an overwhelming number of Basotho, whose forces also threatened their family and farms back in the Orange Free State. So their incursion fizzled out somewhat ignominiously.

An altogether more serious invasion came in 1865. This time a much stronger Boer army under Commandant Fick invested the mountain in the period between July and September. Fick had 2 000 men, as well as a body of Barolong and Batlokoa levies, and some artillery. Moshoeshoe had perhaps an equal number of men on the mountain and a few obsolete cannon.

Fick (who formed a laager of 300 wagons) was joined by Louw Wepener with men from the south of the Orange Free State. On 8 May an assault was made on the mountain. A feint one way achieved little and the main attempt on the Rahebe Pass was undertaken with little enthusiasm. Eight men did manage to scale the cliffs to the west of the pass, thereby becoming, according to Major Tylden's lively account, 'the only fighting men to ever reach the top of Thaba-Bosiu'. But they could not consolidate their position and, after two were wounded, had to scramble down again.

When the president of the Orange Free State, Brand, arrived he ordered another attack on the 15th. The buildings of the Thaba-Bosiu mission station, stoutly defended by the Basotho, were taken by the Barolong, and Wepener commenced the assault on the Khubelu. An artillery bombardment tried to soften up the defenders, commanded by Masupha and Molapo, but it had only marginal effect. The pass was defended by a series of three walls and the attackers had to pick their way carefully upwards, suffering the occasional casualty as they did so. By late afternoon the two lower walls had been taken but Fick found it very difficult to persuade volunteers to reinforce Wepener above.

Behind the last wall, amongst many others, lay defenders Selebalo and Machabo, Mojapitsi and Hantsi Ramathlapani. If they could hold out until the fast approaching darkness arrived, the mountain might be saved, at least for that day. Below them was a large flat rock: onto it in the gathering gloom climbed a bearded man clearly of some importance. They fired a loose volley at him. Hantsi is generally credited with the kill (in 1933 his family, in need of money, sold his famous musket in Maseru). A 'coloured' man jumped onto the rock in an attempt to rescue his commander; he, too, was shot, almost certainly by Mojapitsi, who had probably held his fire in the first volley.

So died Louw Wepener and his *agterryer*. This might have discouraged Fick but it did not: he persisted with the siege, though it did leak like a sieve, allowing visits to Moshoeshoe by the Roman Catholic priest Father Gérard and by a British officer called Reed, who represented Governor Wodehouse and who stayed for several days and enjoyed the hospitality of the defenders, who lived in caves as shelter from the intermittent bombardment.

Eventually a peace was cobbled together as a pragmatic move to enable both sides to complete their harvests – this was known as the Peace of the Millet. War broke out again the next year and, at last, the Basotho nation was brought to its knees. All their mountain fortresses, except Thaba-Bosiu, were captured and the kingdom was on the point of collapse, only their frail king holding things together (Molapo, for instance, had sued for peace with the Boers).

Wodehouse stepped in as a *deus ex machina* in March 1868 and declared 'Basutoland' British territory. This ensured a measure of independence to the Basotho under a relatively benign if largely negligent imperial administration. In 1870 Moshoeshoe who, for so long, had seemed as rocklike as Thaba-Bosiu itself, finally died. While he lived he said, 'This mountain is my mother, had it not been for her, you would have found this country entirely without inhabitants.' He was buried on top, at last permanently within her embrace.

Masupha believed that Thaba-Bosiu now belonged within his sphere of influence and he strengthened its fortifications before, during and after the Gun War, which began in 1880 and petered out in 1883. Although educated at Zonnebloem College, the elite college for the sons of chiefs in Cape Town, and baptised as David by the Paris missionaries, Masupha led the conservative elements in Basotho society and represented the recalcitrant face of the past, never recognising the British authority and snubbing the rule of Letsie and his successor Lerotholi whenever he got the chance.

It is therefore fitting that the last open battle that Thaba-Bosiu witnessed involved this stubborn old nationalist. He was constantly threatening civil war against the reigning monarchy and stirring up mini civil wars.

Finally, in 1897, the crunch came. The British actually considered bringing in regular troops to put an end to Masupha's intransigence once and for all. Instead they urged Lerotholi to take decisive action. In early January 1898 Lerotholi assembled 10 000 fighting men and marched on Thaba-Bosiu. The number of Masupha's followers pretty well matched that of the paramount chief's. Surprisingly Masupha chose not to defend the mountain but drew up his forces to the north and north-east with the main concentration round Khamolane plateau.

On 5 January Letsie made a move on this strategic point but he was driven back. On the following day Lerotholi risked his entire force in an all-out attack. On the left his son Griffith powered up the sides of Berea plateau and pushed back Masupha's right flank. His centre also collapsed. Casualties were light but the battle was decisive. 'It had been an old fashioned action with mounted men moving rapidly against each other and on the left Griffith had led the main charge himself, jumping his horse into a cattle kraal right among the defenders.'

Masupha had to abandon his traditional territory, centred on Thaba-Bosiu and Qiloane. He died the following year far from his homeland. The man who had much to do with him in his last 15 years (and to whom Masupha was a real pain in the neck) was the PEMS missionary Édouard Jacottet, who was the missionary at the Thaba-Bosiu mission from 1885 to 1907. Masupha, he believed, had died of a broken heart. The old man's body was brought back to Thaba-Bosiu and buried on top alongside his father and many other prominent Basotho. 'This death,' wrote Jacottet, 'marked an important date in the history of the country. It is the last vestige of the old Lesotho which has now vanished.'

The end of Masupha, the end of the nineteenth century: it was pretty well the end of Thaba-Bosiu, too. After Moshoeshoe's death the monarchy moved its epicentre to Matsieng, some 40 kilometres away and the administrative and eventual parliamentary activities functioned in Maseru. So throughout the twentieth century Thaba-Bosiu quietly declined into a state of dignified neglect.

Today there are relics of the past that can still be seen. Below the Khubelu Pass can be ferreted out the foundations of the old mission station (started by Casalis in 1836) which was destroyed by a cyclone in 1957. In a small storeroom attached to the new church is a stone said to be the seat Moshoeshoe used when he came down from the mountain (the key to the room can be obtained from a wonderful old lady living in the church house). Spectacular views of Qiloane, Berea plateau and the Phuthiatsana River, witnesses to the passing parade of human frailty, are a bonus. The climax of a visit is, of course, a walk up Khubelu Pass to the top where the prestige of Lesotho lies, in the shape of the chiefly graves, in benign neglect. The prospects from the top are magnificent and there are curiosities to remind the hiker of the romantic past: on the cliff face opposite Qiloane there is the outline of a human foot to which several stories are attached. One says that it was carved by a son of Moshoeshoe who was forbidden by his father to marry the girl he loved because she was not of suitable status: Maleleka was his name and he placed his own foot into the carved mark and leapt to his death on the unyielding rocks below.

As you drive towards Thaba-Bosiu on the tarred road that leads off from the Roma road, the top of Thaba-Bosiu appears for a moment above a green curved slope of a nearer hill. Nothing spectacular. Then it disappears. As you round the next curve a full view of the mountain begins to emerge.

The Mountain of Night has a presence. Towards evening it becomes almost magical. The shades of the past. A host of dramatic events, triumphs and failures, re-enact themselves in the active and receptive mind. The spirits of Arbousset and Casalis, of Masupha and Molapo, of Louw Wepener and Chinese Gordon (martyr of Khartoum) and above all of Moshoeshoe, seem to move through the rocks which come alive in the fantastic fading light.

All Basotho, perhaps all South Africans, should see Thaba-Bosiu before they die. Along with the better-known Great Zimbabwe and Isandlwana, it is one of the three great sites of the southern part of the African continent.

CHAPTER TWENTY-THREE

—•◆•—

TAFELKOP

M otorways are the muzak of roads. With everything reduced to monotone, they numb the mind into neutral and airbrush all detail out of the surroundings. For the dull parts of the journey, they whisper seductively, speed shortens the boredom.

The millions of people who drive the N3 between Johannesburg and Durban every year would probably say that the 100 kilometre stretch between Villiers and Warden is the most boring road in the world. They would be wrong.

Right next to the motorway, on the axis between Vrede and Frankfort, stands a flat-topped hill. It is easily recognised since there is nothing else like it in the area. Not surprisingly, it is called Tafelkop. It has another name, too, which is not on most maps.

There is an exit right there from the motorway which gives easy access to a good view of it, if one takes the trouble to pause. It is certainly worth a pause, but let us defer the pleasure for a while and explore the hill's neighbourhood.

Motorways should be used creatively. You can leave them and join them later because the small roads are where real life takes place. To savour the approach to Tafelkop and if you are approaching it from Johannesburg, take the alternate route, the R103, the old main road to Durban. Along this road – the old road of Death but now a pleasant country ramble – is the typical platteland town of Cornelia.

A detour is worthwhile. The town was founded in 1918 and named after the wife of the last president of the Orange Free State. Its attractions are modest but interesting: aside from a rather forlorn monument to the founding of the Republic in 1961, there are, in the grounds of the church, wagon ruts set in concrete from the 1938 commemorative trek, the 1949 Voortrekker monument celebrations and the 1952 Van Riebeeckfees. In the main street there's an interesting cheese factory and in a side street the observatory which a Pretoria astronomer used for many years, away from the lights of the Reef where the air is clear. Cornelia was the birthplace of the Afrikaans Baptist Church (founded by Odendaal) and the buildings are still there. It is also the birthplace of the writer P.G. du Plessis, who wrote the play *Siener in die Suburbs*. The fine old rectory next to the church was where his father, a teacher, lived, though P.G. himself was born in a humbler dwelling on the corner of Lourens and Odendaal streets, opposite the Sonskyn Kafee.

Out of town there is a tarred road to Tafelkop but a direct dirt road halves the distance.

For centuries Tafelkop has had an almost mystical significance.

In 1836 the French missionary Thomas Arbousset, together with a colleague, François Daumas, made a pioneering journey northwards from their mission base at Morija in Lesotho. They were looking for the source of the great rivers of southern Africa – the Senqu (Orange), the Lekoa (Vaal), the Namahadi (Elands or Wilge) and the Tugela. They mistakenly thought they had found a common source and, being typically French, they gave the mountain a French name, 'Mont-aux-Sources'. In the course of the journey they travelled through the area we know today as the eastern Free State. So devastated by constant war had the area become after the depredations of Shaka that Arbousset called them 'the fields of the dead'. All that was left of many villages were the skeletons of the cattle kraals which the locals would simply call *lerako* (walls).

Towards the end of April Arbousset nearly drowned crossing the Namahadi when his horse was thrown flat against some rocks by the force of the current. He then made his way towards the unmissable flat-topped hill.

Three streams run away from Tafelkop. Arbousset followed the middle one known as the Noka Tlou or Little Elephant River (now called Rietspruit) upstream to the hill which the Basotho called Ntsoanatsatsi (the rising of the sun). Two hundred years before, the plateau on top of Ntsoanatsatsi was bordered by wild olive trees, found nowhere else in the vicinity. At the foot of the east side of the hill was a Bafokeng village but its chief held court under the trees at the top.

Long, long ago, Chief Napo of the Bakwena moved there from the west and married a daughter of the local chief. Napo was to become the great-great-great-great-great-greatgrandfather of the greatest of the nineteenth-century chiefs, Moshoeshoe. Generations later, after they had moved south to Lesotho, the Basotho still believed that most of their clans had come from Ntsoanatsatsi.

So much so that Arbousset, when he visited the hill, took note of a Basotho myth that there was a cavern there, surrounded with marsh reeds and mud, from which they all (that is, mankind) sprang. He also noted that the tribes turned the faces of the dead to the east when they buried them. Apparently, as far away as Villiers, graves have been found facing Tafelkop. For centuries, it was clearly a special place.

This mythical significance is by no means the only thing worthy of note. The hill was also the centre of considerable action during the Anglo-Boer War, with several battles in the vicinity, little-known but worth a visit.

About 7 kilometres outside of Frankfort lies Aanleg Guest Farm, where I stumbled across a superb guide to the area. Although a busy farmer, Bertie de Jager took the time and trouble to share his knowledge generously with me. Close to his farm, on 11 October 1900, there occurred a sharp little encounter which epitomises the sadness and savagery of the war as effectively as much greater and better-known battles like Spioenkop or Magersfontein. A small Boer force of about 54 men, keeping an eye on the garrison at Frankfort, took up a position in an isolated *poort* concealed from the Vrede-Frankfort road. They posted a couple of vedettes – one at Jakkalslaagte on Aanleg Farm – and erected two large round tents for their comfort in the *poort*.

In an unusual move for that time, the British sent out a force towards evening. Next morning, before daybreak, they were in position within spitting distance of the Boer camp. Surprise was complete. Christaan Beyers, who got up early to see to the horses, suddenly found himself in the crossfire of the British ambush. Danie Cronje, from Vrede, was deaf so he was not woken by the opening shots: he died instantly when a bullet took him in his sleep. The shooting lasted no more than five minutes. By the time the Boers surrendered they had lost 11, including Veldkornet de Wet and two pairs of fathers and sons. The wounded and captured were taken to Frankfort, and most were ultimately sent to St Helena.

The skirmish is not mentioned in the *Times History* of the war and the site can be found only by someone with intimate local knowledge. Only a small plaque marks the general area but, when the wind blows through the grass in the *poort*, it moves the imagination to try to cope with the idea of fathers and sons perishing together and a man dying without ever hearing the shot that killed him.

It was believed that the Boers' position was betrayed by a black servant called Stuurman, so that the deaths were seen as murder. Not recorded on maps, the site is known as Moordpoortjie.

Towards the end of 1901 the British began to build a line of blockhouses between Frankfort and Vrede (as an extension of the line from the west) in order to pen the Boers in south of the line. They were all of the Rice-type design. The structures of corrugated iron and stone have all disappeared now; only the circular foundations crouch low in the grass.

De Jager is a quiet, almost shy man. But, paradoxically, those who speak most slowly are often those who have most to say.

He has completed the Comrades Marathon ten times. When he is feeling lazy he gets his wife to drive him more than 20 kilometres into the veld and he is forced to run home. So he knows every inch of the countryside and has actively sought out and pinpointed 54 blockhouse sites. He is the only one who knows where most of them are. One site is high on a hill on his own farm and from original plans he has lovingly reconstructed it so that it is a landmark easily visible just to the north of the R34 Vrede–Frankfort road.

As you sit up there, replete and benign from a braai that the Tommies who manned it a hundred years ago, sick of bully beef, could only have dreamt of, the 360 degree view is magnificent. Down below, bordered with green willows, the Wilge winds its way, a paradise for canoeists, if only they knew it.

Leading away from the blockhouse is a short stretch of original barbed wire that joined the forts, which were spaced only a kilometre or so apart. The wire never ran directly between the two forts. It was always in a V-shape so that if a sound was heard at night, the defenders could shoot down its length without threatening their compatriots in the next post.

Over to the east, the now familiar Tafelkop rises implacably out of the undulating plain. By mid-December 1901, the line of blockhouses extended 19 kilometres out of Frankfort but was still 13 kilometres short of the hill. The British learnt that a Boer force of some 300 men under General Wessel Wessels was somewhere to the south, so, on 20 December, they sent a force under colonels Rimington and Damant to sweep around the east side of Tafelkop. Unfortunately, their two columns got separated as they moved south-westward.

Damant's forces, following the Rietspruit with their two artillery pieces, a Pom-pom and a Maxim, reached a hill with a magnificent all-round view. Down towards the river they saw the Boers firing at a group in distinctive British uniforms. Too late it was discovered that this was a trick – the latter were actually Boers in disguise. The British cannons were stranded in open ground, their crews annihilated. Damant himself was shot four times. Though he survived, 33 British soldiers were killed and 45 wounded. The *Times History* knows it as the Battle of Tafelkop but the site is actually some kilometres off, the mountain itself standing majestically to one side like an unblinking sentry. The battle site is not easy to find but it is worth the effort to do so as the scenery is superb.

Again, the best person to guide you is De Jager. He can point out the characteristic circular foundations of blockhouses unseen before your very eyes; he can conjure the vestiges of trenches out of bare veld.

Eventually Tafelkop itself was ringed with six blockhouses, and three artillery pieces were positioned on its top. The line of fortification moved on to the fine little town of Vrede, where local resident Willem Naudé will show you that the handsome church follows the design of Notre Dame in miniature. If you are very lucky you will get him to talk about the horrifying battle of Kalkkrans (or Langverwacht) which took place to the south of Vrede on the night of 23 February 1902.

Initially, the Boers had scoffed at the potential effectiveness of the fortified line.

Now, they had come to realise it had the thin deadliness of a garrotte. At Kalkkrans, De Wet, with 2 000 fighting men but also many aged and women in his train, as well as President Steyn and his staff, proposed to break out in order to make for the Witkoppies to the south-east. They started a night attack and by midnight, in the teeth of murderous fire, De Wet, Steyn and 600 men did manage to extricate themselves. But the vast majority did not, and had to retreat – to be mopped up as prisoners in the days and weeks ahead.

The war is a living thing in the area, passed on in oral tradition. There are, too, some monuments: in Vrede are the graves of 23 New Zealanders killed at Langverwacht; in Frankfort, some of those who died at Moordpoortjie are buried alongside the 166 British casualties in that district.

There are also more personal memorabilia: the farmer at the battle of Tafelkop site has the revolver his grandfather took off Damant after he shot him; De Jager will show you the two holes made by the bullet which passed through the hat his great-grandfather was wearing at Moordpoortjie (he survived); Naudé has a military map of the area stained with the blood of the British officer from which it was taken. Small reminders of a war that should never have happened.

From Vrede let us return to Tafelkop one last time to think about an association which is perhaps the most interesting of all.

Arbousset and his colleagues Casalis and Gosselin, of the Paris Evangelical Missionary Society, were, in 1833, the first missionaries to enter Lesotho. They established the mission initially at Morija. In the decades to come the mission spread throughout the country, and flourished. Morija, with its printing works, became the centre for the publication of almost all books in Sesotho. It also published a newspaper, *Leselinyana*, which is today the oldest surviving African newspaper, more than 130 years old.

In the early part of this century, at the Book Depot which distributed Morija's publications, worked a young clerk called Thomas Mofolo, whose book, *Moeti oa Bochabelo* (Traveller of the East), *Leselinyana* began to publish in 1907. It was the first novel written by a black African anywhere in Africa. It is a work of enormous power.

In it, the hero, Fekisi, grows dissatisfied with the dissolute life he sees around him in his village. He asks the elders about God. They tell him that He *does* exist but lives far away, beyond a place called Ntsoanatsatsi, which stood in a great reed bed surrounded by much water from which a fountain rose. Man himself, they said, came from the reed bed.

One night Fekisi has a dream: he sees Ntsoanatsatsi in its reed bed and surrounded by its marsh, and the figure of a man passes before him in a transparent mist. The man's face is so beautiful that Fekisi cannot look at it, and he weeps. So the fictional character of Fekisi, drawn by vision, follows the path of Arbousset to Ntsoanatsatsi.

After he leaves it, the journey becomes much more allegorical and Fekisi travels to the sea, where he is taken by three white men into a ship and travels to a sanctified death in the City of God across the ocean.

Tafelkop is therefore more than a one-stop on the road to Durban. It is associated with a creation myth like Adam and Eve; it is the centre of courage and sacrifice, of betrayal and deception and the inevitable signatures of war; and it is an inspirational point in the birth of the African novel.

All this is less than two hours' drive from Johannesburg, right under our noses. The trip can be done in a day; but it deserves a relaxed weekend. And, as local tourists, it demands our support, even our veneration.

But you who have to travel east in a hurry spare a thought, pilgrim, for the place where the sun rises because you are about to follow, for the zillionth time, the way that Mofolo's hero took towards the City of God.

If, on the other hand, you are coming up from Durban, make your peace with the place since, after you pass it, you will soon be in the City of Hell.

Whichever way you come, however, do not say the place is boring, or that there is nothing there.

CHAPTER TWENTY-FOUR

—•—

NAKOB

Nakob is not exactly a household name. Children the length and breadth of the land can seldom be heard clamouring: 'Daddy, daddy, take us for the weekend to Nakob.' In fact, it's a safe bet that 99 per cent of the population do not know where the hell Nakob is, and it would make a good 'Trivial Pursuit' question.

Yet an incident happened there where not a shot was fired and perhaps no more than 200 metres of territory were involved but it helped precipitate South Africa into one of the biggest wars it has ever entered.

Nakob is situated on the border between Namibia and South Africa on the main road between Karasburg and Upington. Now it has a brand-new red-brick border post. In 1914 it was a small two-roomed wood and galvanised iron building, a couple of dilapidated tents and just about the remotest part of the country, 'remarkable only for wide, void spaces'. To the locals it was known as 'Groendoorn' and was situated on the farm Aries.

The countryside to the south is rugged and broken, a hodge-podge of gullies, gorges and mountains, called the Noup Hills. In the early part of the century it had provided hiding for the Nama hunted by the Germans from South-West Africa and had witnessed many a deadly skirmish.

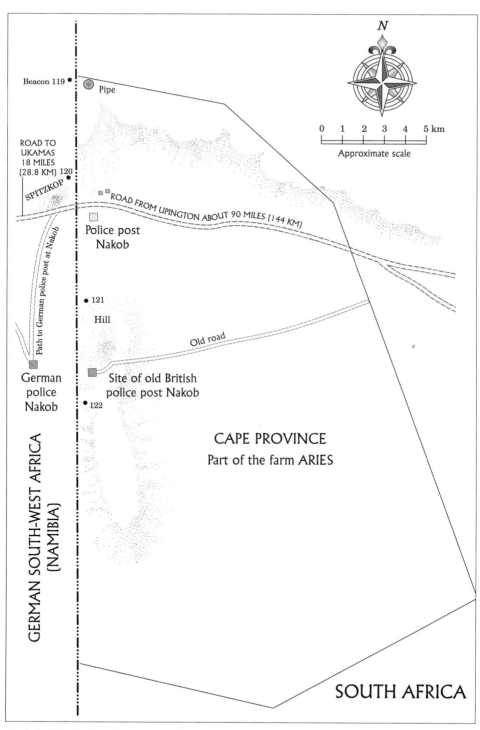

Nakob 1914 (after Fred Cornell's map)

Snaking to the west is the Gariep (Orange) River, cutting into the rock and allowing only a few crossing places – Schmit's Drift, Raman's Drift, Viool's Drift and Zendling's Drift, close to the sea. Often the banks are so sheer that no access is possible and men have died of thirst within sight of water. Up and down this land, for many years, Fred Cornell operated as a diamond prospector. He knew and cherished its contours like a well-tried lover. He enshrined this passion in a wonderful book called *The Glamour of Prospecting*.

The boundary between the two countries runs in a dead straight line along the 20th degree of longitude east. In 1914 it was marked by a set of international beacons spaced a few kilometres apart with the British coat-of-arms on one side and the German on the other. The police post at Nakob was usually manned by three men of the South African Mounted Rifles (SAMR); 80 kilometres to the north was a small post called Obopogorop and 80 kilometres beyond that was Rietfontein, the seat of the magistracy and a detachment of police whose mounts were camels.

Today the modern tarred road sweeps past a few places of interest only to those in the know – Toeslaan, Lutzputs, Langklip, Cnydas, and the Brak River. (About 40 kilometres from Nakob a long low flat mesa which Cornell called 'the escarpment' appears to the north of the road, while to the south lies more open sandveld.) Ninety years ago to get to Nakob was a much more difficult exercise. Eight hundred kilometres of rail to De Aar from Cape Town, then another 160 to Prieska might take only 36 hours but then truly atrocious roads would make the 240 kilometre trip to Upington a nightmare.

Then there was still the 130 kilometres from Upington to Nakob, either over rough ground strewn with sharp stones or over sand 'so fine as to resemble snow', which sometimes meant that in some cases trekking might progress 'only a few yards in a few hours'. In some parts of the border area the adventurer would need one of the locally bred horses which 'could climb like cats, live on stones and sand, and go wherever a man could go'. Take a wrong path, however, and one could wander lost until death.

Wasteland Nakob may have been but it did have water and Fred Cornell decided to buy a farm adjacent to the South African border post. A near-death accident – a terrible fall from his horse which smashed ribs and an arm and left him lying helpless in the desert – kept Cornell in hospital in Upington until 30 July, so he only arrived at his camp a couple of kilometres north of Nakob late on the night of 4 August 1914.

Only next morning did he hear that war had broken out between Germany and Russia. The three policemen at the post were Corporal James Hall (a young Sussex man), Trooper Green (a fine shot), and Trooper Human, a handsome Afrikaner. They had not got the news from Upington – communication on the South African side came via a 'galloper', a man who ran two days on foot to deliver a message! They got it from the new 'enemy', the Germans, who, at Ukamas (18 miles from the border), had telegraph and telephone.

War or no war, routine had to be adhered to. Nakob was an important trade route and usually a large stack of lucerne was piled up there waiting for the German wagons to fetch it. But at this time of year there was no lucerne, only 2 000 salted skins – known locally as 'Gordonia Bank Notes' – waiting to be collected.

Hall and Green set off to deliver a prisoner to Upington (a task which required two days as a minimum) so Human had to consult Cornell on the legal phraseology of a document, brought in by a galloper, warning that war was imminent. But it was only on 8 August that they heard that Britain (and, nearly simultaneously, South Africa) had declared war on Germany on the 4th. South Africa's declaration of war had the serious effect of imperilling the internal security situation. Many military officers were unhappy at the prospect, decided on by Prime Minister Louis Botha on 6 August, that German South-West Africa would be invaded; many important Afrikaner leaders felt that this was the time to reverse the result of the Anglo-Boer War and throw off the jukskei of British rule.

By 8 August there were five SAMR troopers at Nakob. They were in a tense situation. At Ukamas the Germans had 200 soldiers backed by Maxim guns and the latest Mauser rifles with their nickel-pointed bullets, sharp as a lead pencil, whose effects on human flesh were terrible. On the border the Germans replaced the few policemen, whom Cornell respected, with 25 military troopers, whom he did not.

In fact, Cornell had little liking for the Germans. He had visited Ukamas where he had witnessed their humiliating behaviour towards the lone Jewish storekeeper there. He frequently made reference, too, to what he called 'the holocaust at Shark Island'. During the Herero 'rebellion' in 1906 hundreds of Herero, many in chains, naked and wounded, were landed on that small, half-submerged islet in Lüderitz Bay and left there for months, packed tightly like penguins, unsheltered and uncared for, scorched by day, and chilled by the sea fog and spray at night. Hundreds died of hunger and thirst, dozens were driven mad by drinking salt water in their desperation; scores drowned trying to reach the mainland; many killed each other fighting over scraps.

For years Cornell had been concerned over the loyalties of the 'poor neglected ignorant' farmers of the northern Cape, neglected by their government and who often had nowhere to turn for the necessities of life and for medical truth but to the Germans across the border. And the Germans had 'insidiously attempted to sow the seeds of rebellion' in their turn. He was aware that German money had long been the common currency on the streets of Kakamas and in the wilds of Gordonia and that English silver was rarely seen. Indeed, he believed that the abortive raid of Ferreira in 1906 was openly instigated by the Germans as a *ballon d'essai* to test the way the wind blew.

The precarious position of the isolated SAMR post was becoming increasingly apparent and snippets of intelligence filtered in on 12 August indicated that it was about to be attacked. The place was impossible to hold – it was on a flat plain commanded by hills all round. The most prominent feature in the landscape was just across the border on the German side – a conical hill called the Spitzkop.

(To this day it looms formidably over the new border post.) So at night, after lights were out, the five troopers slipped away into the mountains to the north and joined Cornell and his two companions, an Englishman called Ford-Smith and an Afrikaner wagon-driver called Jan Nickel, in his camp.

The troopers were not the only ones doing the slipping. Cornell's labourers began to melt away. Every day one or other of the small garrison climbed the escarpment with its clear view for many miles along the road to Upington looking for the signs of approaching reinforcements. None came.

Cornell began to think no further useful purpose would be served by staying on. But he was a self-confessed spy, if only an amateur one. He he¹ become accustomed, 'simply out of sheer love', to bringing in reports oᵢ ᵢs various trips to Major Buckle, intelligence officer at the Castle in Cape Town. So before he left Nakob he decided he would try to ascertain the whereabouts and strength of the Germans. On the morning of the 15th, having finalised his preparations for departure, he took Hall and Trooper Bartman to a high point from which the country could be scanned within a 20 mile radius. In a direct line south (some 4 miles from where they were standing and a couple of miles south of the South African border post) stood a prominent and isolated granite koppie, studded with huge boulders. It was two to three hundred foot high and was situated 'slightly on the British side of the line'. The original South African border post had been sited at its foot before the post was moved to the north closer to the water supply but it still marked the locality of the German police post, which was placed, out of sight, to the west of it.

Cornell had often passed by 'this bold hill' but had never climbed it. It must, he thought, overlook the German post; conversely, it would provide a commanding view over South African territory to the east. He decided, before leaving, to take a closer look at it.

At 3.30 that afternoon he called at the SAMR and was 'rather hipped' that he could get no one to accompany him. Armed only with a camera with just three 'films' left and a small shaving mirror to signal with, he made his way cautiously towards his target, using the fairly thick melkbos as cover (I have tried this myself and think it works!).

Having made his way safely to the base of the hill past beacon Number 121 he cautiously worked his way up it. Halfway to the top he found, to his astonishment, a well-beaten track for horses and a newly made loopholed *schantz* overlooking South African territory. Open-mouthed with surprise he carefully made his way to the top where, surrounded by titanic boulders, there was a flat space roughly the size of a tennis court. It gave 'every sign of recent occupation' – paths trampled by many feet, more *schantzes*, a gun emplacement, a hammer for breaking stones, even a still-burning camp oven. Recently used bottles lay scattered about. All were within South African territory.

Over a ridge to the west he could see the German flag above the border post and at the base of the hill he spied a man in khaki, sitting, apparently waiting for relief horses. Cornell quickly snapped the *schantzes* with his remaining three photographs and, taking advantage of the fading light, crept away, though not without a last glimpse of two silhouettes standing on the top of the hill he had just vacated. The Germans, he now knew, would have ample warning of any relieving force coming from Upington or Kakamas.

Cornell could not leave till morning and the five young troopers spent the night 'laying on their arms' near his wagon. Their supplies were nearly exhausted, the wagon with their monthly rations being long overdue, so Cornell left them what little he could spare before he lumbered off on his own heavy wagon. When he was beyond the hills of the Brak River – out of sight of the watchful eyes of the German intruders on that big granite koppie in British territory, which 'should have been held by British troops as soon as possible after war was declared' – he blew up all his dynamite 'just to give the Germans something to think about'. He left Ford-Smith with the wagon and rode on alone.

On the first day he met not a single soul and spent the night at Toeslaan (where the main road is intersected by the road to Lutzputs) alongside a huge fire, roasting and freezing by turns, since it was bone-chillingly cold. The rough rocks and heavy sand ensured that he and his pony only arrived, dead-beat, in Upington at midnight the following day, neither of them having eaten anything.

Next morning, at daybreak on 20 August, he woke up Captain Fisher, officer commanding the local SAMR, and told him of the situation. Fisher took down his sworn statement and a telegram was sent to Pretoria with the news that 'German troops had invaded British territory'. The press seized on this and the prime minister used it sensationally to back the (contested and controversial) decision to join the war and ultimately invade German South-West Africa.

Meanwhile Fisher told Cornell that there was no provision for sending reinforcements to Nakob and he felt he needed specific instructions from Pretoria. In any case, Upington, Cornell was aware, was 'a lukewarm place – not over conspicuous for loyalty'.

The timing of the Nakob incident was critical. On 15 August a large meeting had been held at Treurfontein in a district where rebellious feeling had been stirred amongst Afrikaners by the prophecies of the *siener* Niklaas van Rensburg. Though General de la Rey had there counselled calm, many of the old Boer generals were secretly plotting. A meeting of military officers discussed the campaign against the Germans on the 21st. An incident at Schmit's Drift (in which a German patrol pursued a renegade family of farmers called the Liebenbergs across the border into South Africa) added further outrage to the violation of Union territory. The three-pronged invasion plan of German South-West was bedevilled by the suspect loyalty of Lieutenant-Colonel Solomon Gerhardus 'Manie' Maritz, commander of the Union Defence Force in the Gordonia area, and the plotters regarded a meeting with Maritz as critical to the success of a rebellion. Over the next few weeks Maritz played a duplicitous game.

On 9 September at a special session of Parliament Botha used these incidents of trespass to back a resolution 'fully recognizing the obligations of the Union as a portion of the British Empire'. On 18 September South African troops landed in Lüderitz Bay and soon after elements of General Lukin's force were moving overland into the area south of Warmbad.

Cornell had once met Manie Maritz. He found him to be an alert, bluff, soldierly man, of medium height and well groomed, with good English and an educated manner. Sturdily built, he nevertheless gave little sign of the enormous strength he was known to possess. Whatever his faults, Cornell saw him as no hypocrite – 'he never professed to have any other feeling than that of hearty detestation for the English'.

Subsequently, though, Cornell had little but contempt for him: 'I only know too well how the treacherous skunk denied the truth of my own statement.' Cornell believed that Maritz deliberately deceived Smuts and Botha: '... can it be doubted that his lying statements as to the position in general on the North-West border lulled the authorities into a sense of false security in that quarter, and prevented them from taking measures to protect this remote, forgotten portion of the Union from invasion and rebellion?'

The post at Nakob was consequently never reinforced in the weeks that followed. Early on 16 September, Andries de Wet approached the small building with a force of over 200 men. It was the first operation of the Vrij Korps (Afrikaners living in or exiled in German South-West), but included a number of Germans with two maxims. They sent a message demanding the surrender of Corporal Coulter and his six men. Coulter refused. After a second demand was turned down the Germans opened fire. Coulter fell just as he was about to mount his horse. The handsome Human had his jaw shot away, making a dash for liberty.

The rest jumped into the shanty and flung themselves flat on the floor but the firing, including the maxims, continued without respite. Corporal Hall, now in charge, decided they could not hold out, so Rifleman Ignatius Botes hoisted a bed sheet out of the door and the shooting stopped. Coulter was lying outside, shot through the back, dying. The others were sent as prisoners to Ukamas.

Through his treachery in 'discrediting any news that came from loyal sources' and in commanding 'the troops that should have protected them' but didn't, Maritz had, in Cornell's eyes, 'by means of his traitorous lying tongue and his German gold' betrayed his own men at Nakob.

Today, if you are passing through the border post on your way to Keetmanshoop and Windhoek, take a look north to where Cornell had his camp, to the west at the unmistakeable Spitzkop, and to the south to the isolated granite koppie. Spare a thought for Coulter, spare a thought for Human. But, above all, consider the incident where not a shot was fired in August and where only a couple of hundred metres of territory were occupied, but where an opening bid in a bitter war was made.

And you might not forget Nakob in a hurry.

CHAPTER TWENTY-FIVE

KEIMOES, KAKAMAS AND UPINGTON

Those travelling to the battlefields of Keimoes and Upington from Johannesburg or Pretoria can smack their lips and whet their appetites on the long road there by visiting a few sites, like trailers to the main feature.

In Kuruman not only are both the Eye and the Moffat mission a definite attraction but a brief detour northwards up Livingstone road which runs into Seodin road will bring you to an old dead tree called The Silent Witness (or The Truce Tree). In November 1914, General Christoffel Greyling Kemp made his way westwards from the Transvaal with his group of 700 rebellious burghers in order to link up with General Manie Maritz in Gordonia and with the Germans of South-West Africa. Kuruman lay in their way.

Captain J.G. Frylinck was under orders from the Union government to defend the town and he gathered a disparate force of soldiers, policemen and coloured citizens to perform the task. A skirmish took place at Pakhane but Kemp deemed it too costly to capture Kuruman and since this was not his primary purpose – which was to garner supplies and pass through – he induced Frylinck to meet him for talks. This meeting took place under the camelthorn tree on 8 November. Frylinck refused to surrender government arms, horses and property but was allowed to leave Kuruman without being fired on. He joined up with Commandant van Zyl and they followed Kemp, harrying him like pesky dogs, and bringing him to battle at Mamaghodi on 13 November and across the Langberg at Witzand on 16 November (where Kemp lost four men killed and eleven captured).

As you approach Olifantshoek the Langberg come into view from many kilometres away. The Langberg are precisely that – long, low hills which seem to present from afar not particularly formidable terrain. But, when in dire straits, the Batlhaping retreated there as a refuge of last resort and it was there in 1897, under chiefs Galeshiwe and Luka Jantjie, together with Toto, the Batlharo chief, that they did make their last stand against colonial forces determined to eradicate them. And as you get close to the hills you realise the waterless, jagged, stony outcrops present a nasty obstacle to any attacking force.

On the outskirts of the village of Olifantshoek a minor dirt road (marked D3332) leads northwards along the eastern side of the Langberg. Twenty-four kilometres along the road one reaches the farm Gamasep (where there was formerly a police station). Here, on 30 July, a government force of 2 000 men, supported by heavy artillery, broke the back of the 'rebellion'. Luka himself was determined not to submit and confronted a section of the Kaffrarian Rifles at point-blank range. He was gunned down and a captain of the Cape Town Highlanders commissioned a surgeon to cut his head off and boil it. It was subsequently hauled around in a sack.

On the farm there is a small cemetery which you will not find without the help of the owner, Pieter van Wijk, who hospitably took me to it. It is hidden away in the thorny veld. The most legible of the graves is a Jewish one, containing the body of Lieutenant Mark Harris, son of Ephraim and Rebecca Harris of Manchester, who died in battle on 8 April 1897 aged 28 years. In the hills around are the remnants of old fortifications and cartridge cases can be found in the sand.

On the way out of Olifantshoek the D3331 leads a short way to the turn-off to Fuller Farm where there are three graves of men who were killed in a war of 1878 and that is a story in itself.

But we must hurry on, for Gordonia is our main destination. 'The Green Kalahari', that miracle of irrigation and industry all along the Orange River, is a very pleasant place indeed to spend a few days exploring.

When war was declared between Germany and Britain in 1914 many Boers saw this as an opportunity to take up where they had left off at the end of the Second Freedom War (the Anglo-Boer War of 1899 to 1902). When Prime Minister Botha declared war there were protests in parts of the country and a plot to rebel was hatched on General Christiaan de Wet's farm Allanvale outside Memel in the eastern Free State. Amongst the ringleaders were General C.F. Beyers, head of the Union Defence Force in the Transvaal, Jopie Fourie, an officer of the Defence Force (who was court-martialled and shot on 20 December 1914) and, perhaps, General Jacobus Hercules de la Rey, a senator and former hero of the Anglo-Boer War. The rebellion was given quasi-spiritual and prophetic sanction by the seer Niklaas van Rensburg. In Upington, Lieutenant-Colonel Manie Maritz, commander of District 12 (which included the districts of Kuruman, Prieska, Gordonia, Kenhardt, Clanwilliam, Calvinia, Van Rhynsdorp and Namaqualand) of the Union Defence Force, was expected to join the rebellion, timed for 10 October.

Almost immediately after the outbreak of the Great War, Britain requested the South African government to undertake a military expedition against German South-West Africa with the immediate aim of capturing the wireless stations at Lüderitzbucht and Swakopmund, which were of great assistance to the German sea raiders in the south Atlantic. This the prime minister, General Louis Botha, and the minister of defence, General Jan Smuts, agreed to do and a session of Parliament, meeting first on 9 September, endorsed the move on 14 September with a large majority.

Some Afrikaners, under the leadership of generals Hertzog, De Wet, De la Rey and Beyers (at the time still head of the Defence Force), protested against the proposed invasion, maintaining they would only defend the Union's borders, although by some of them and their followers, a full-scale rebellion was plotted. The incidents at Nakob and Schmit's Drift somewhat undermined their position, however, and support began to cohere behind the Union government, even amongst some Afrikaners.

At the 21 August meeting of military commanders in Pretoria a plan for the conquest of German South-West Africa began to take shape. General Sir Duncan Mackenzie and 'C' force would invade from the sea, Colonel Lukin and 'A' force would enter through Port Nolloth and cross the southern border of German South-West Africa at Raman's Drift, and Maritz was to mobilise 1 000 men from the North-West Districts as 'B' force and operate in concert with Lukin, perhaps through Schmidt's Drift or Nakob.

A two-week training camp for units of the Active Citizens' Force was held at Potchefstroom starting on 2 September under the instruction of Major Kemp. It is clear that the plotters intended to use these regiments as the nucleus of the rebellion's forces in the Transvaal. Kemp resigned from the Defence Force on 15 September (so that he would not be guilty of treason), and he was followed by Beyers.

About 8 o'clock on the night of the 15th Beyers and De la Rey left Pretoria by car and headed for Potchefstroom via Johannesburg. At 9 o'clock they came across a police roadblock ironically looking not for them but for the notorious Foster gang! When their car did not stop after being challenged a warning shot aimed at a tyre ricochetted off the road and struck De la Rey in the back, killing him within a few minutes. Many saw his death as a government conspiracy (almost certainly mistakenly) but the start of the rebellion was perforce postponed.

'At Upington,' in the words of one of the government's blue books on the Rebellion, 'Maritz was waiting for the signal to begin' and the arrival of Beyers and De la Rey at Potchefstroom Camp 'would be the signal of revolt'. The plan was for Seitz, the governor of German South-West, to come across to Raman's Drift with

2 000 men and Maritz would start by attacking Upington.

Maritz had been active in the Second Freedom War and, as an opponent of surrender, had subsequently trekked to German South-West Africa where he'd played a part, on the German side, in the Herero Rebellion. After he returned to South Africa he was accepted into the Defence Force and was appointed as Staff Officer of Military District No. 12 at the end of July 1914. He was virulently anti-British and during the run-up to the Rebellion he had secret contacts with the Germans.

With the shooting of De la Rey 'the golden moment had passed' and, in the depressed days that followed for the conspirators, Kemp remarked to a friend, 'Thank God, we've still got Manie Maritz on the Orange River.'

On 26 September a column of Union forces was badly cut up by the Germans at Sandfontein, south of Warmbad in German South-West Africa, but Lukin persisted with his plans to advance on Warmbad. There followed a dangerous dance between Maritz and Pretoria who instructed him to move part of his force to Schmit's Drift. Maritz pleaded that the Germans had 3 000 men at Ukamas with artillery while his own troops were untrained. He telegrammed saying that he had 'only three machine guns under two little English lieutenants who seem to me to be scarcely able to fasten their breeches with the few children under them'. He added that 'I will do my best to support you on this side of the border'.

Even though they suspected Maritz of playing a double game Botha and Smuts could not yet replace or dismiss him for fear of provoking him. Instead, they appointed Major Enslin, nominally as his chief of staff, but in reality as his watcher. They also deployed the Imperial Light Horse, a mounted regiment, and the Durban Light Infantry, an infantry battalion, to Upington. These arrived on 4 October and were put under the command of the popular loyalist Colonel Coenraad Brits.

Alarmed by their approach Maritz moved his forces closer to the German border to Wilhelm Frank's farm, Van Rooisvlei, 25 miles west of Upington, and closer to Kakamas. On or about 7 October he met the head of the Schutztruppe, Colonel van Heydebreck, commander of the German forces, at Ukamas and obtained his support. Andries de Wet with the Vrij Korps and a German force of four pom-poms and two machine guns under Colonal Hausdenk were sent to link up with Maritz at Van Rooisvlei (they arrived on 10 October).

Meanwhile Brits had sent a letter to Maritz instructing Maritz to report to him in Upington on 9 October but Maritz prevaricated. Brits thought Maritz wanted to provoke him into attacking with English troops, thus allowing him to make political capital out of the action and set a civil war going. He sent Major Bouwer to Van Rooisvlei with a letter instructing Maritz to hand over his command to Bouwer. But on 9 October, the day before Bouwer arrived, Maritz had at last made a decisive move, perhaps because he had injudiciously revealed the secret plans of the rebellion to a garrulous acquaintance.

Maritz surrounded and disarmed the 50 men of the Maxim gun section under lieutenants Freer and Botha (they were mainly from the midland and eastern districts of the Cape and consequently less likely to join him). He then called the whole camp together into a circle round him as he stood under a camelthorn tree and announced openly for the first time his intention to go into rebellion. The Vierkleur was hoisted and he gave the men one minute to decide whether or not they would join him. About 60 refused so they were disarmed, arrested and taken to the German border where they were handed over as prisoners. The majority (about 800), driven by their commander's powerful personality, went along with him (though over 100, indicating that they were reluctant participants, took the earliest opportunity to desert).

Consequently, when the unfortunate Bouwer arrived, he was arrested and told he was 'a prisoner of the German Empire'. When Bouwer said to him he must be joking, Maritz replied: 'No, I am not. Yesterday, I went into rebellion against the Government. I asked for my discharge long ago and the Government would not grant it to me. Now they expect me to go to German South West and take that country with a lot of boys, which is impossible for me to do. They will die of starvation and thirst, and therefore I am obliged to take the steps I have taken.' Bouwer might have pointed out that he had taken a lot of boys into a treasonous rebellion but, perhaps tactfully, did not do so. Instead, he was sent back to Upington bearing the message of Maritz's defiance. On the basis of Bouwer's report Martial Law was proclaimed throughout the Union on 12 October. In the letter sent with him, Maritz told Brits that: 'I love my country and my people, and I do not wish to be the cause of bloodshed between Afrikander and Afrikander.'

Brits' retort on the next day was stinging: 'I can assure you that the majority of the Dutch-speaking people support the Government … You have no monopoly of patriotism. The blood and tears of hundreds and thousands of Afrikanders which will flow in this war which you have brought about and declared, be on your head. By your action you have placed our people in danger of being destroyed as a nation, both by the enemy and through fratricide and parricide … You know of what crime each ignorant child under you has been guilty. You know the punishment to which they are liable. Nothing on earth can justify such an act. Posterity will judge as to who was in the right – the man who fulfilled his duty as a citizen, or the men who used their position as officers and men of the Union Defence Forces to deliver their brother officers and men to the enemy.' He denied Maritz's allegation that soldiers were forced to serve across the border: 'It was not intended to compel any person to go across the border against his will … It is not now a question of going across the border; that the Government has done long ago. There are already 120 young Afrikanders lying shot dead in German South-West Africa.' It was beginning to dawn on Maritz that this was not going to be a simple English versus Afrikaner war. It was, in fact, a civil war.

Instead of attacking Upington, as was planned for 11 October, Maritz decided to retire to Kakamas in order to be closer to the German border. Here a Proclamation was issued establishing a Republic. But there were daily desertions and the German officers regarded the position at Kakamas as unsatisfactory. They forthrightly told Maritz he must either fight or give up his plans. So when Maritz heard that Union troops under Captain van Rooyen had arrived on the outskirts of Keimoes he took a commando there and set up his headquarters in the post office while his men were largely encamped around where the Spar shop is now. Many of the townsfolk hastily abandoned the place, while the rebels dug themselves in.

The battle itself took place on 22 October and can be seen from two fine vantage points. The place to start is the small hills behind the De Werf chalets guest house on the way into Keimoes from Upington. Fortified by some excellent crumbed pork chops I made my way up one of these hills, with small jagged sharp rocks underfoot, making the going slightly difficult. The hill is overrun with dassies and still offers up cartridge cases to the sharp eye. At the top are a couple of broken-down and partly dispersed *schanzes*. The Union troops under the command of Johann Leipoldt initially took up their positions here.

From this hill one can easily see the terrain of the battle. The Orange River is to the left and the town of Keimoes is in front straggling northwards along Van Rooisvlei. Beyond the town and across the valley are a longish row of hills where the Vrij Korps (or Freikorps) and the German artillery had taken up their stance as backup to Maritz.

The most prominent feature in the landscape is a clump of large hills to the left across the river, called the Oranje-berge. The highest is Tierberg. To prevent being surrounded Maritz decided to take the initiative. Some of the German artillery seems to have moved to the low hills around what is now the railway station and wine cellar, from where they began pounding Leipoldt's position, forcing him to retreat across the river via Keimoeseiland and link up with reinforcements from Upington.

At this point it is worth transferring yourself to the top of Tierberg (you can't miss it – it's got KEIMOES written on it). A short winding road through the Tierberg Nature Reserve brings you to a breathtaking view. Below is the small town of Keimoes and its surroundings like a great green oasis (it would not have been quite so extensive or lush in 1914). Again one can view the layout of the battle, this time from the second dominating position of the government troops. Now Leipoldt's Hills are in front but to the right; the town and Van Rooisvlei stretch out directly in front, with the hills marking the German artillery position to the left.

In the great scheme of things this was a minor scuffle but Maritz, watching the battle astride his horse Appel, took a bullet below the knee. He had to leave the battlefield to be patched up and his small army retreated in the direction of Kakamas under cover of the artillery. The battle did therefore bring home to the rebels that the government was no pushover and that a great spontaneous rebellion may not have been on the cards.

Brits and his troops followed up the retreat and again attacked the rebels under Commandant Stadler at Kakamas on 24 October. The Vrij Korps and the Germans

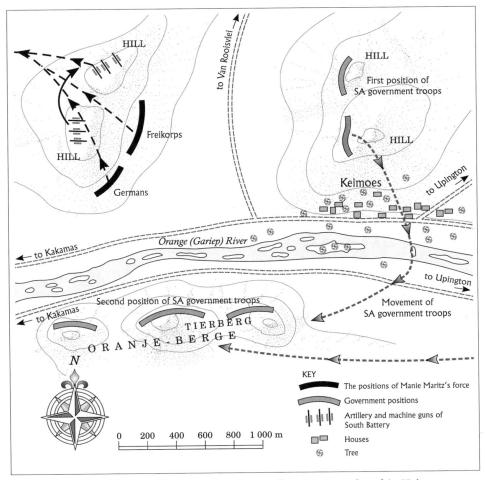

The Battle of Keimos (after an original German drawing reproduced in Keimos en Omgewing *by Maria de Beer)*

with their artillery retired over the border, taking Maritz with them. Maritz and his surviving men licked their wounds in a place near Schmit's Drift called Jerusalem.

An interesting sideshow to the actions in Gordonia was the expedition of about 200 men (under Major Ben Coetzer) that Maritz sent southwards in order to stir up support for the cause. They walked with ease into Kenhardt, whose inhabitants had fled, then, further south on a farm called Bakke, divided their force, one column headed for Calvinia, the other for Carnarvon. This latter column, which included two Maxims under Lieutenant van Weiherr, a German officer, encountered a loyalist scouting party, under Major Naude, at a place called Breekkerrie on 24 October. An officer was sent under a white flag to demand Naude's surrender. To his surprise he met his uncle and several friends and discovered, contrary to what Maritz had assured them, that the district was not in a state of rebellion. The consequence was that the whole rebel party surrendered instead! The column directed to Calvinia was captured two days later at a place called Onderstedoorns on the Zak River by a small force of Major P. Vermaas. This ended Maritz's incursions south of the Orange.

While Maritz was licking his wounds at Jerusalem, Kemp was fighting all the way westwards from Vleeskraal and Schweizer-Reneke. Exhausted, with clothes ragged, their feet blistered, their boots disintegrated and their horses worn out, Kemp and his commando eventually reached Zwartmodder on 26 November where they linked up with a patrol of the Vrij Korps. They were hotly pursued by a strong force from Upington and a small group of Maritz's men had to fight a rearguard action to protect them. This rearguard eventually stopped the pursuit at a farm called Koegoekop (to the southwest of Zwartmodder). One of the rebels in the party went by the *nom de guerre* of Kalahari Mac. Forty years later he wrote a lively account of his experiences in a book called *Agter die Skerms met die Rebelle*. Through his eyes we can taste the dust and danger of the campaign.

Freddie MacDonald ran a few cattle and sheep on the farm Kraskop in the district of Kakamas-north as the war loomed. But not all Boers were farmers and not even a wife and three children could tie him down. He had, in his own words, 'avontuurlus'. He fitted much better into that tradition of border adventurers, restless and opportunistic, who had roamed the area for decades and of whom Scotty Smith (now buried in Upington) was the brightest star.

MacDonald knew the countryside better than almost anyone (he had fought in what he called the 'Second Freedom War' and alongside the Germans in their Herero War) so it was natural for Maritz in August and September to use him on two occasions to make secret contact with the Germans in South-West.

According to MacDonald's own account Maritz sent him on his first mission to persuade the Germans not to violate South African territory again as they had done in the first Nakob and Schmit's Drift incidents since this gave propaganda ammunition to the Union government. The second mission came after the firefight at Nakob on 16 September when MacDonald was to persuade the German commander, Van Hydebreck, to rein in Andries de Wet and the Vrij Korps. Having delivered this second message MacDonald attached himself to the Vrij Korps and saw action at Sandfontein, where the invading advance column of Unionist troops, foolishly clustered around the single eponymous hill, was given a nasty bloody nose and the bulk of them were compelled to surrender in a situation which one of them described as being 'worse than hell itself'.

When the September uprising was aborted Kalahari Mac found himself in a tricky position – technically fighting alongside his government's sworn enemy. He did, however, make one scouting trip southwards but when the Rebellion did finally break out he transferred to the eastern sector. When Maritz and his dwindling force limped into German territory after the battle at Keimoes, MacDonald with a heavy heart joined him at Jerusalem. He estimated Maritz had barely 300 men, augmented by the mere 300 of the Vrij Korps. Morale was low and doubts over Maritz's competence, like insidious gangrene, set in.

Von Heydebreck wanted to launch a combined operation into Gordonia; Maritz felt this would merely drive the majority of Afrikaners into the Union fold. Von Heydebreck wanted to use an iron fist; Maritz preferred the use of gloves. In the end MacDonald felt the German commander was vindicated. An attack in September or even November might have worked: by January it was too late.

When Kemp and his men struggled into Jerusalem and the nearby farm of Blydeverwag, they brought with them Siener van Rensburg, whom they transparently revered. He was short, with a long black bush of a beard, a stately demeanour and a Bible always under his arm. He never smiled.

MacDonald often had dreams himself, which warned him of approaching danger. Van Rensburg's visions were usually enigmatic, their interpretations contentious, but MacDonald and his compatriots were a superstitious lot and usually the Siener's prophecies proved to be accurate in their eyes. No one who met him, said MacDonald, doubted Van Rensburg's sincerity. But when Van Rensburg, fully conscious, described a vision of a bunch of donkeys chasing some springbok, and then the roles being reversed, what on earth was one to make of it? Kemp, however, found he could not move without the prophet's say-so.

Before the arrival of Kemp, Maritz had been, despite the promptings of the Germans, pretty timid. After, there was some rivalry between the two generals as to seniority and the two could not agree on strategy. Maritz wanted to invade Namaqualand but even his own officers were vehemently against the plan: Namaqualand was thinly populated and there were only two routes for retreat (through Goodhouse or Raman's Drift) which the Union might easily cut off. Maritz made only one half-hearted encroachment into enemy terrain, surprising the defenders (on 18 December) at Nous. He captured 130 prisoners but soon retreated once again beyond the Orange.

The Germans at this stage received a serious setback. They had two light aircraft and improvised some bombs for them. When Von Heydebreck and some of his officers went to inspect them one of these bombs fell, exploded and the capable commander was killed. He was replaced by the eccentric, monocled alcoholic, Major Franke. It was only after New Year that the rebels began to move on the eastern front, though with little German backup (had they allowed the Germans a significant presence in the campaign the results might have been different).

The Maritz-Kemp force was about 1 200 strong and Freddie MacDonald was in the forefront as a scout as they passed through Nakob. Because of the scarcity of water the force had to advance in small groups. At Langklip, 30 miles in, MacDonald's small patrol was approached by a lone rider who seemed unaware of their existence until only a mile away when he suddenly turned and fled. MacDonald and others gave chase but he escaped. When MacDonald later found out this was the coloured 'spy' Jan van Wyk, he kicked himself for not having pursued harder. (There was an 'agreement' that neither side would use 'coloureds' and any coloured participant captured was usually executed, as indeed was Van

Wyk when he was finally caught – summarily, at close range, through the head.)

From prisoners the rebels learnt that Kakamas and Upington were both strongly defended so their whole force was concentrated on Lutzeputs (or Lutzputs) where there was a large encampment of Union mounted soldiers. Lutzeputs itself (12 kilometres south of the main road to Nakob at the Toeslaan turn-off) is situated in open terrain but with a range of hills to the south. The rebels surrounded the Union soldiers and threatened them with artillery – a large group was captured and disarmed but a couple of hundred found a weak point in the encirclement. There ensued a pell-mell flight across the plain towards and into the rough and rising ground, with MacDonald and his men streaming after them. Pursued and pursuing horsemen strung out in a long line – those with the weakest horses were hauled in and made prisoner. Mile after mile the chase continued, until they often became individual duels. MacDonald singled out one and slowly caught up with him. When he got close he stopped, dismounted and shot the fleeing horse, taking care to miss the horseman. The trooper took refuge behind his dead mount and fired a shot as two of MacDonald's men joined him.

It was a classic Boer scenario two centuries old, starting with the Bushmen. No massed ranks of warriors, no red-jacketed squares, no impersonal artillery or scything maxim. Man against man. Simple, almost elemental.

The three men riddled the dead horse which could provide no shelter against the Mauser bullets. '*Toemaar, kêrels*,' said MacDonald, standing up, believing the incident to be over. A second shot came from the direction of the dead horse. The three rebels, joined now by two mates, immediately turned it into a sieve. This time no one got up to check, but a defiant shot rang out once more from behind the dead animal, which must have been lying over a hollow depression so that the rebel bullets were going through it but too high!

While the sniper was kept preoccupied with concentrated fire MacDonald trotted over and saw the man lying on his back behind the horse. Macdonald barked an order, '*Staan op.*' But the man remained flat as a pancake, saying, 'I can't understand.' MacDonald repeated his command in English. 'I can't. I'm wounded,' was the reply. On closer examination MacDonald observed that it was a light wound '*deur sy sitplek*'. 'It's not serious,' said MacDonald, 'I've also taken a bullet in the same place,' and offered the man a hand up. 'Oh, yes,' said the Englishman, 'I can stand quite well.'

He was an out-and-out rooinek. When MacDonald berated him for carrying on fighting when cornered, he replied, 'We're all here for the same game and I'm merely doing my duty.' MacDonald told him he could thank his lucky stars he was still alive whereupon the Englishman said, 'I'm just here to die for England.' MacDonald could only laugh and offer his hand, 'That is brave of you. I like a brave man.'

As the man returned the handshake and expressed appreciation for his capturer's conduct, MacDonald smelt alcohol. So he wondered whether it was the alcohol or the man who was so brave.

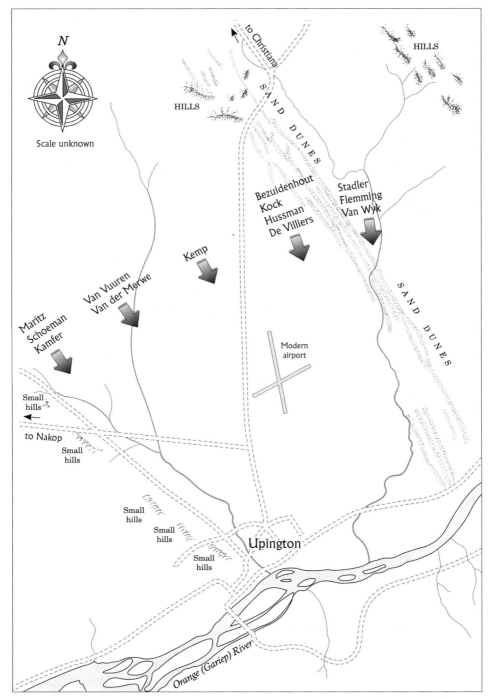

The Battle of Upington (after a drawing by A.K. Cornelissen in his monograph, Langs Grootrivier*)*

For Kalahari Mac, Lutzeputs was child's play compared with Upington. Because of the difficulties in water supply the rebels advanced along a line to the north of the main road, through Grondneus and Smalvis. Their target was the farm Christiana, which is 25 kilometres to the north of Upington. The sand made the going heavy for the wagons, and it was easy to get lost. Siener van Rensburg went with them on his donkey-cart. He was asked about the future. 'Yes, I can see us standing in rows in Upington, receiving food and sugar. I've never been in Upington and don't know how it looks. It's funny, but I can't see the road out of the town. Perhaps I'll see it later.'

MacDonald took heart. There was no doubt that they would capture Upington, otherwise how else would they enter the town since Van Rensburg had seen them there, and the seer's foretellings were always dead sure.

Christiana is a very pleasant farm almost surrounded by a graceful sweep of ridges and the rebel commandos took up scattered positions on them. Between Christiana and Upington there is a large grey hill which the rebels next took. It is only 13 kilometres from Upington and has a beautiful view of the town.

Kemp and most of the officers wanted a pincer movement skirting the town and reaching the river above and below it so that the attackers could take advantage of the cover the dense foliage of the river provided. The contrary Maritz never took direct advice and he ordered a frontal assault.

The attack began at daybreak on 24 January, along a front of five miles. The 2 000 strong garrison under General van Deventer were well dug in and the rebels had to advance over bare ground. Maritz's main thrust was delivered against Van Deventer's right wing. As the light improved the fighting became more severe and there were heavy casualties on both sides. A small fort and a cattle kraal were captured by the rebels but they were later dislodged from them. At 11 o'clock the defenders brought up two guns which compelled the attackers to withdraw their own artillery.

Kalahari Mac (who was on the west side of the action) was full of praise for the bravery of the Transvaal rebels, unwaveringly trusting in their prophet. But the situation was hopeless. Seventy rebels were captured and the rest began the slow withdrawal towards Grondneus.

Flight or capitulation was all that was left. On 2 and 3 February most of the 1 200-odd rebels, including Kemp, surrendered in Upington. Maritz, who had helped arrange the surrender, fled to South-West Africa. So did Freddie MacDonald. He seems to have been involved in a surprise attack on Kakamas, which was easily repulsed (the graves of seven German soldiers lie against the hill at Lutzburg outside the town), and he later helped the Germans suppress the Reheboth uprising. When German resistance finally collapsed in South-West, he took refuge in Angola.

But the seer had been right after all: the road of the Rebellion did end in Upington. It was only the interpretation which was slightly off the mark.

CHAPTER TWENTY-SIX

— • • —

POTTER'S HILL

On one of the district roads between Memel in the north-east Free State and Charlestown in KwaZulu-Natal a lone monument stands on the southern slope of Majuba mountain. The inscription translated from Afrikaans does not give very much away.

This memorial
is erected
in memory of
Fannie Knight
and
G. Roets
Who were shot by S. Swart
on 6 May 1927
While he had escaped
from the police
at Potter's Hill.

Enigmatic. Mysterious. But, boy, what a story lies behind it.

There are any number of ghost stories associated with the small town of Memel. In fact, every second house seems to have its resident restless spirit.

In the house of Hennie van der Berg the apparition is a lady who walks in the night. On the farm Negefontein, where old Muir was murdered, heavy footsteps tread the passage. Nearby you might be startled by the lights of a ghostly car which appear and as quickly disappear, and lead you on to some ghastly fate. Then there is the doctor who, while playing snooker in the hotel, felt his pulse and announced that he would die that night. He went home, shaved and died. Apparently, he walks still in his white surgical overalls.

But the most spectacular tale is the true story of Fanie Swart. It is a story where the 'Gunfight at the OK Corral' meets 'High Noon'.

Stephanus Swart, origins unknown, arrived in Charlestown and Volksrust, its twin town just across the border in the Transvaal, in 1920 with a batch of stallions for sale. Since he had had previous convictions for stock theft, prevailing opinion was that these horses might also be of dubious provenance.

One story has it that when he sold a horse to the widow Annie Eksteen, coins rolled out of her purse like golden dice. With an eye for the main chance he soon married her: he was in his mid-thirties, she was over sixty. Her main attraction was the farm Potter's Hill, a 'mountain paradise' up against the looming presence of Majuba. She in turn got a fine big stallion of a man with, according to hindsight, an evil cast to his face and *snaakse oë*.

Like the mountain next to which he took up residence, Swart was an angry man, brooding, subject to frequent violent outbursts and unprovoked quarrels.

One of these fights was with his stepson-in-law Willie Knight whose homestead was close by. Swart resented Knight because he rented the farm Short's Cliff from his mother-in-law, Annie – Swart's wife – for only a pound a month. So Swart 'brutally thrashed' Knight to the point where Knight had to draw a revolver to protect himself and Swart got 18 months in prison for assault. During his incarceration he hardly spoke, not even responding to the most solicitous enquiries from his fellow inmates. He nursed a deep sense of grievance against his wife's daughter, Fannie Knight, for giving evidence against him on her husband's behalf.

Every man's hand, he considered, was against him. To his workers he was a fiend of gross magnitude. But his special hatred was reserved for the officers of the law: for the magistrates, who, he felt, were in a conspiracy against him, and the police, who, in his eyes, persecuted him remorselessly and wanted to break him like a wild horse.

He, on the other hand, had an unswerving faith in his own rectitude: he was, he believed, above the law.

The only things that he loved were his animals: his Arabian stallion, his Afrikander cattle, his watchdogs. When he caught one of his workers surreptitiously kicking one of the dogs he went for his gun and the worker only survived by disappearing over the horizon.

He spent hours at a time staring at Majuba. Often he climbed it, alone with himself for long periods. He had so much inchoate and undefined anger and frustration within him: only the mountain was big enough to stand up to him.

He regarded himself, however, as the embodiment of manly sexuality. Most women feared and avoided him but many were attracted to him and he flaunted his affairs in front of his wife. It was this which was to precipitate the events in May 1927.

Two months before, he had locked his wife and his nephew-manager, Alwyn Visser, in a dark room and threatened their lives. Swart was having a relationship with a 17-year-old relative whom he had brought to live in the house and incest charges were imminent. (One version still current has it that the girl got pregnant and Swart tried to persuade a neighbour Roets to acknowledge paternity. Roets refused.) Visser escaped with his life only by lying and promising not to give evidence against Swart. Mrs Swart took flight to Potchefstroom.

The charge of *bloedskande* (incest) was duly pressed against Swart and witnesses like Visser and Mrs Swart were brought to Volksrust, but Swart ignored the summons. The die was cast.

On Tuesday, 3 May, Swart went to sort out Knight once and for all but Knight, forewarned, fled, so Swart took a shot at the farm manager, Roets, instead, fortunately missing.

Swart then initiated a series of actions which indicate that he was burning his bridges. He drove his motor car to Newcastle and on the way back set it alight and walked home. He shot his beloved dogs, perhaps because they might betray him. Then, on the evening of the 4th, he summoned his lawyer to the farm and, over several nerve-wracking hours for the lawyer, dictated to Mr Maasdorp a 28-page final testament.

The last words, directed at his numerous enemies, including two chief justices, were chilling: 'I now give blood for blood. I will shoot them down until I have one cartridge left, and that will be mine; but alive you will never get me. With my corpse you can do what you please. Burn it, mutilate it, and treat it in such a manner as you think you can best revenge yourselves.' He concluded the interview in the early hours next morning by declaring, 'I am going to stop all traffic over my farm from 6 p.m. tonight.'

A law unto himself, Swan effectively turned Potter's Hill into an independent country.

Presented as they were with incest, contempt of court and the attempted murder of Roets, and informed of Swart's state of mind by Maasdorp, the police felt compelled to act. The responsibility fell on the district commandant at Volksrust, Captain Gerald Ashman.

Ashman was born in Somerset, England, in 1872 and had come to Bloemfontein after the South African War. He joined the Orange River Colony municipal police, then the CID, and promotion was rapid. He was married with three children.

He was joined by the head constable of Newcastle, William Mitchell, a popular and fearless Irishman, daringly reckless in the interests of justice.

In all, 12 policemen formed the posse, and they were accompanied by Alwyn Visser who volunteered (not without some hesitation) because he knew the layout and terrain of Potter's Hill. Visser warned them about his uncle's marksmanship. Swart could throw a bottle in the air and then, from 50 paces, with a pistol in his left hand and a Mauser in his right, fire simultaneously and both bullets would hit!

At 4.30 a.m. on 6 May, they assembled in the lounge of the Belgravia Hotel in Charlestown where they were given breakfast by the proprietor, A.E. Lloyd, and his wife, who also provided them with sandwiches for later in the day. Ashman was aware of the toughness of the task ahead and requested that the papers in his room be given to his son if anything unforeseen should happen. When Lloyd asked him if they had any bandages he replied that he did not want to alarm his men but would take them if they were provided.

The road crawled painfully over the shoulder of Majuba and, after a torturous drive, the posse arrived at the boundary of Potter's Hill in two motor cars and a motor-cycle with sidecar at about 6 o'clock. Just inside the fence was an Indian store which was about three-quarters of a mile from the farmhouse. Here they paused to plan strategy. One of Swart's workers, carrying a note to his *inamorata* in town, was stopped, the note confiscated and the horse tied to the fence. The police divided into two, one party worked its way uphill, the second downhill. The idea was to approach in extended formation and surround the farmhouse. For the moment it seems that Ashman and Sergeant Annes van Wyk remained at the store, perhaps to co-ordinate the offensive.

The mist was down, visibility was restricted to a few metres, the men moved like spectres through the cold air and the smallest noise startled the ear. It was a cold and unfriendly morning on which to die.

There was some dispute about what happened first. Constable William Feucht's version was that, as he approached the kraal, he saw a man moving amongst the sheep. As he neared the kraal corner he saw Swart coming out of the gate carrying a Mauser rifle. Feucht called out, 'Hands up!' but Swart ran back into the kraal without a word. Just then, Feucht saw 'a native' hidden amongst the mealies some 20 yards away, who fired and Feucht fell. He was dreadfully wounded, by ricochet, in the left eye, arm and thigh, which indicated to some that a shotgun had been used.

Afterwards the authorities denied that any African workers had fired a shot and no shotgun was produced. More than likely it was Swart in the mealie field and he was master of the situation from the time the first shot was fired.

Feucht was carried back to the Indian store and reinforcements were requested. Swart was now regarded as *voëlvry*. Mitchell and William Crossman, a 23-year-old first-class constable born in the eastern Cape, now crept up to the corner of the kraal wall, thinking Swart was there.

He wasn't. He was in the field behind them. And he was close. His first shot wounded Crossman. Mitchell dashed for cover but Swart shot him down. Swart then walked up to them both and dispatched them in cold blood. He then reloaded, leaving a little pile of Mauser shells next to their bodies.

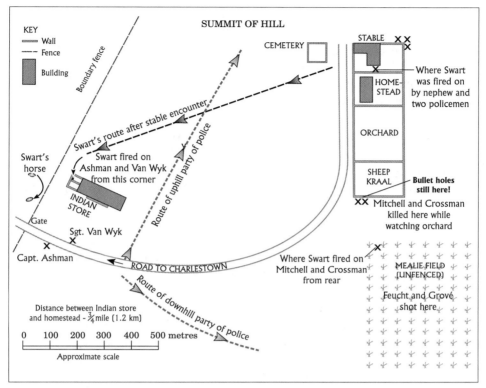

KEY
- ═══ Wall
- --- Fence
- ▮ Building

SUMMIT OF HILL

CEMETERY

STABLE XX X

Where Swart was fired on by nephew and two policemen

HOME-STEAD

ORCHARD

SHEEP KRAAL — Bullet holes still here!

XX Mitchell and Crossman killed here while watching orchard

Boundary fence

Swart's route after stable encounter

Route of uphill party of police

Swart fired on Ashman and Van Wyk from this corner

Swart's horse

INDIAN STORE

Gate

Sgt. Van Wyk

Capt. Ashman

ROAD TO CHARLESTOWN

Route of downhill party of police

Where Swart fired on Mitchell and Crossman from rear

MEALIE FIELD (UNFENCED)

Feucht and Grové shot here

Distance between Indian store and homestead - ¾ mile (1.2 km)

0   100   200   300   400   500 metres

Approximate scale

Gunfight at Potter's Hill (after a contemporary drawing in The Star*)*

Sergeant Grové was 30 and had a wife and two children. He was the expert tracker and finest shot of the party. He began to stalk Swart through the mealie field in this deadly cat-and-mouse game. But in the mist hunter became hunted. Grové was shot from behind, completely outstalked by Swart. The bullet penetrated the ribs to explode two of the rounds Grové kept in one of the front pouches of his bandolier. Grové crawled away into the mealies.

Swart now wanted to get away so he made for the stable to get a horse. But the uphill party, which included Sergeant Moorcroft and Alwyn Visser, spotted him and challenged him. Momentarily confused by his khaki overcoat they momentarily held fire and he escaped. The belief was that he was now in the stable: in fact he was behind it and he worked his way down to the store, using it as a sight-shield.

At the store Ashman and Van Wyk were supervising the evacuation of Feucht in one of the vehicles (after Van Wyk had dressed his wounds). As the vehicle left Ashman walked alongside it giving the driver last-minute instructions.

But, completely unknown to them, though seen by Hans Kumalo, a farmworker, and Jan Jacu working in his own mealie fields behind the store, Swart was now lying behind a low wall to the side of the store. Firing deliberately, he shot

Van Wyk as he stood in the road. Ashman returned fire, getting off two shots, before Swart gunned him down easily. Ashman pitched onto his hands and knees and started crawling away. Realising the significance of the retreating car Swart fired at it, too, but without effect.

Swart then strolled up to the prostrate Van Wyk and put a bullet in his head and another in his heart. With Ashman he didn't bother. Ashman was already dead. But Swart did take his Webley revolver.

Providence sidled up to the brave. Swart's horse, which Ashman had ordered hitched to the fence, was waiting for him. He mounted it and rode away in the direction of Charlestown.

At about eight o'clock he approached the farm Oude Spruit of his neighbour, S.J.M. Swanepoel. 'I've just shot five policemen dead,' were Swart's first words. 'The Lord has been good to me today.'

'Don't be afraid,' he told Swanepoel, 'I shall not harm you.' He then asked for a cup of coffee, drank it, shook hands and rode away. As he did so his last words were, 'Oh yes, tell them they will find Grové's body in the mealie field. If you do not, they will not find it quickly.' It was about 12 miles to Charlestown.

Swanepoel immediately took off for Mount Prospect, a small village to the south of Charlestown, to warn the police, riding past O'Neill's cottage where the truce for the first Anglo-Boer War was agreed. As he was nearing the village he got a note from an African sent by Swart saying he would return to Swanepoel's farm to shoot him. Swart had apparently changed his mind!

The news of the bloodbath had reached Charlestown (first from the car carrying Feucht to Volksrust) and its inhabitants were panicking. Fannie Knight was worried about her children and her husband back at their home near Potter's Hill. She persuaded Roets to drive her there in a trap. Mrs Lloyd tried to dissuade her, telling her it was not a wise move.

A black called Somsewar, living in a hut 50 yards from the road on the southern slope of Majuba, watched as Mrs Knight's trap came over the ridge and down the road. In the other direction he saw Swart galloping very fast up the road and passing in front of his house. Seeing the cart, Swart got off his horse and strode towards it, gun in hand. With the other hand he waved the trap down. He said to Roets, 'Are you living with Mr Lourens?' Then he fired at Mrs Knight and she slumped over the side of the cart. After that he shot Roets, who collapsed dead in the road. The monument today marks the spot where they died.

Mrs Swart was staying in Charlestown with people called Van Vuuren while she waited to give evidence in the incest trial. The Van Vuurens lived in a small wood and corrugated iron house behind the single main street and 300 yards from the station. When he rode into town this was the first place Swart headed for (he had scoured the town for her in the days before and had only discovered her whereabouts on the previous day).

On the verandah Lukas van Vuuren, a 21-year-old cripple, sat with his 17-year-old sister Gertrude. He said, 'Who is this riding so furiously down the road?' Gertie replied, 'It's old Swart,' and dashed through the house, crying a warning to Mrs Swart, before heading for the police station.

Lukas, in the valley of the shadow of Swart's gun, cowered in his chair. 'Where is my wife?' asked the man from Potter's Hill. 'She was here,' answered Lukas from the depth of his seat. Swart said nothing more and stumped into the house. Lukas heard Mrs Swart speak calmly. Then there were two shots so Lukas hobbled off into the next-door yard, 20 metres away, and watched as Swart, 'with a strange set expression in his eyes' and staring without blinking at the distant hills, reloaded on the verandah before walking out into the sunshine and mounting his horse.

Mrs Swart was in the dining-room with wounds in the chest and in the centre of her forehead. One of the bullets was buried in a wooden box beyond her and there were splinters of lead all over the floor. She had been shot as she was pleading, in the middle of her final sentence.

As Swart galloped through the street people ran for shelter. Charlestown was a one-horse town and Swart was on the horse.

He had a list and had three more names on it. He made for another house where he called for Mrs Lourens (against whose husband Swart nursed some unknown grudge) but she seized her child and fled, followed by three shots which missed.

Volksrust was a two-horse town and Swart knew that additional police would be coming from there and from Mount Prospect and Newcastle to the south. He decided he needed transport faster than his Arab stallion. So he rode to the bridge between Charlestown and Volksrust, dismounted and posted himself there waiting for the first suitable motor car to come. While he did so he hastily wrote the threatening note to Swanepoel which he gave to a passing black man to deliver.

Mr Hatley, the hotel manager from Volksrust, was the next unlucky victim. He was driving the first car to arrive at the crossroads, with a Mrs Pulford and her child as passengers. Fortunately he had been warned that Swart was in the vicinity and was on his guard when Swart shouted at him to halt, so he accelerated instead. Swart fired through the windscreen and Hatley was hit in both legs and a bullet went through Mrs Pulford's buttocks and lodged in her calf. Happily the child was unscathed and the car got away.

At last, some pursuit of Swart was underway. The Charlestown stationmaster, Kriel, having armed himself, ducked and dived after Swart as he made his way through town. Kriel was joined by Badenhorst, a lieutenant in the Rifle Association, and a man called Laubser. They followed him towards Volksrust. From Volksrust constables Seaward, De Klerk and Jordaan were coming the other way. Swart was caught between the two parties.

As they approached, Badenhorst fired two shots at 700 yards, then two at 600. Kriel fired at 500 yards and as he saw dust spurt up he adjusted his sights to 700 yards. The first shot was close behind Swart so he fired again. Swart looked round and then began to run. As he got off his last shot Kriel saw, through his sights, Swart pitch over and drop into a sluit. Laubser then tried to shoot Swart's horse but missed.

The Keystone Cops from Volksrust were understandably nervous as they approached and got out of their car carefully, Jordaan on the left side using the car as cover, the other two on the open side.

They found Swart lying on top of his rifle in the shallow ditch on the side of the road. He had a revolver in his right hand, an automatic pistol in his overcoat pocket, another in his trousers. His head was in a pool of blood which dyed the green grass dark red. The entry wound of the bullet was through his left temple, the exit wound through the right. He had folded his hat neatly under his head like a pillow.

There was afterwards some dispute about the cause of his death. Was it Kriel who had killed him? The inquest was unequivocal: Swart had shot himself. He had used Ashman's service revolver.

Gradually the extent of the carnage was discovered. Swart had killed eight, then himself (some sources suggest he shot and killed in addition a passing black near Majuba but this did not come up at the inquest, is disputed and needs further evidence). The bodies of the policemen at Potter's Hill were found with the sandwiches from the Belgravia Hotel littered around them. Grové had managed to crawl half a mile before he died amid the alien corn.

The following evening (Saturday) over a 100 people gathered to bury Fannie Knight and Annie Swart at the small cemetery at Potter's Hill. The separate funeral of the five policemen, buried with military honours, was attended by two to three thousand mourners. The body of Captain Ashman was placed on an old gun carriage drawn by a police horse, neighing loudly and decorated with black and white silk trappings. The dead man's boots, reversed, were strapped into the stirrups.

A few days afterwards the area panicked again when a rumour spread that Swart's brother had arrived to take revenge. The authorities had to deny the report.

And what of Fanie Swart? At first nobody wanted anything to do with his body, neither Natal, nor the Orange Free State, nor the Transvaal. There is a point on the S465 district road where the three provinces met: one solution was to bury him there! Eventually he was buried surreptitiously in the police yard at Charlestown. Afterwards some locals regretted this and wanted the body to bury in their own orchards to scare off intruders and to prevent their fruit being stolen.

Many blacks did not believe that Swart had died at all and that his spirit still broods around the mountain of doves – Majuba – and can still be seen moving through the mist, presumably in his khaki overcoat.

Some of the sites can still be seen. If you drive out from Memel on the main A34 road to Newcastle turn left on the S465 about four-and-a-half kilometres outside Memel. As you travel along this dirt road (tricky in wet weather without a 4 x 4) look in passing to your left for a fine old bluestone farmhouse, one of the oldest in the district (it is called Seekoeivleipoort). This has a ghost which blows out candles when there is no draught.

Also on the left (as you pass the Quagga relay station) is a small wood where African National Congress guerrillas, in former troubled times, kept a weapons cache. Ghosts of a different kind.

About 35 kilometres from the turn-off from the main road you will come across on your right-hand side a small yellow sign saying H.F. Adendorf: Potter's Hill. The sheep kraal where Mitchell and Crossman died is immediately to the right: bullet holes in the stones can still be seen if you look carefully. Only the foundations of the old farmhouse and stable survive but the graves of Fannie Knight and Annie Swart are still there, in a small cemetery on the farm. Six kilometres further on, beyond the Mahlambamasoga River, is the monument to Fannie Knight and G. Roets. In Charlestown the five policemen lie quietly side by side at the top of a gentle-sloping hill. Swart's grave is more difficult to find. If you drive past the police station you will see an old corrugated iron building which was the original courthouse. In the heat of this tiny shack (it was built in 1889 and should be a national monument) the inquest was held a fortnight after the shootout. In an open field behind it, 100 metres away, Swart's simple grave is surrounded by low railings and marked with a piece of iron with his name and dates rudely inscribed on it. The grave is difficult to find but worth the effort. (If you come from Volksrust and Charlestown, you take the Quaggasnek turn-off 5 kilometres outside Charlestown and will find the Fannie Knight monument 5 kilometres along the dirt road.)

But the story does not end there.

Years after that dramatic morning in 1927 some Africans in the area of Majuba were celebrating a marriage. They decided to use the monument as a focal point for the wedding photographs. One photo was taken of the bridegroom leaning against the plinth. When the photograph was developed he was seen leaning against ... nothing. Instead, alongside him, was the figure of a white woman, dressed in old-fashioned clothing! The ghost of Fannie Knight.

I have not seen the photograph. I am told that it does exist.

SOURCES

While much information has been gleaned from oral sources and by visiting all the battle sites, some printed sources have been used extensively and should be seen as influences throughout this book. Amongst these are G.M. Theal's *History and Ethnography of Africa south of the Zambesi* (London, 1909), three volumes, and G.M. Theal's *History of South Africa since September 1795* (London, 1908), five volumes; and G.E. Cory, *The Rise of South Africa* (Cape Town, 1965), six volumes. There are different editions of these histories: these are the editions I have used (though occasionally consulting the others).

On the eastern Cape, in particular, there are, of course, many sources. But the standard work (up to 1858) must be Noel Mostert's fine book *Frontiers* (London, 1992) and I owe him an intellectual debt. But some find the book daunting, at least at the start. A happy way into the history of the region is John Milton's eminently readable *The Edges of War* (Cape Town, 1983). It is sadly out of print. More specific sources follow, although they are not a comprehensive bibliography.

Chapter One: Lower Sabi
The main source for the skirmish near Lower Sabi is De Kuyper's journal, published in H. van der Merwe, *Scheepsjournael ende Daghregister* (Pretoria, 1964), pp. 185–209. See also G.M. Theal, *History and Ethnography of South Africa before 1795* (London, 1909), v. 2, pp. 461–480.

Chapter Two: Saldanha Bay (1)
The best book for the history of the Bay is Jose Burman and Stephen Levin, *The Saldanha Bay Story* (Cape Town, 1974), particularly pp. 57–64. Other information comes from R.J. Pieters, 'The Battle of Saldanha Bay, 1781, Viewed against the

Present-Day SADF Principles of War', *Regulus*, v. 4, No. 1, August 1981, pp. 39–47; I.J. van der Waag, 'Naval History of Saldanha', *Paratus*, v. 45, No. 1, January 1994, pp. 45–46; Mary Kuttel, 'Naval Actions at Saldanha Bay' (provenance unknown); C. Truter, *West Coast Tourist Guide* (Cape Town, 1986); and a pamphlet produced by the Saldanha Bay Tourism Office on the history of Saldanha Bay. For the early Khoi history see the article on Saldanha Bay by Ian van der Waag in *Military Academy Annual*, 1996, pp. 68–71. The relevant chapter in Theal, op. cit., is at pp. 123–167.

Chapter Three: Muizenberg
Most information comes from Theal, *History and Ethnography of South Africa before 1795* (London, 1909), v. 3, pp. 237–277. See also, Jose Burman, *The Cape of Good Intent* (Cape Town, 1969); P. Erskine, 'Admiral Elphinstone's Naval Task Force 1795–1796', in *Antiques in South Africa*, 1983, pp. 83–90; and 'East Fort to get Priority of the Trust', in *Hout Bay and Llandudno Heritage Trust Newsletter*, May 2001, pp. 1–6.

Chapter Four: Saldanha Bay (2)
See Chapter Two above. Theal's account of the battle appears in *History of South Africa since September 1795* (London, 1908), pp. 9–16. The relevant section from Burman and Levin's *The Saldanha Bay Story* can be found at pp. 65–75.

Chapter Five: Blaauwberg
The main works consulted were: Jose Burman, *The Cape of Good Intent*, pp. 53–59; Theal, *History of South Africa since September 1795*, pp. 127–150; James Grant (ed.), *British Battles on Land and Sea* (London, n.d.), v. 2, pp. 334–335; W. Brinton, *History of the British Regiments in South Africa* (Cape Town, date unknown) pp. 24–29; R. Cannon (ed.), *Historical Record of the 72nd Foot or the Duke of Albany's own Highlanders* (1848), pp. 37–42; and P. Groves, *History of the 93rd Southern Highlanders* (1895), pp. 3–7. I also encountered several miscellaneous pieces not always attributable or whose authors I could not identify: 'The Battle of Blaauwberg Continues', a 13-page report of the Table View Rangers; a list of British casualties at the battle (source: Major A. Gordon); a one-page typescript entitled 'Historical Sites in the Blaauwberg Area'; a nine-page typescript entitled 'The Battle of Blaauwberg: An Incident in the Napoleonic Wars'; a manuscript letter from S.J. Lee re 'Captain Charles Lee and his son John (Johannes Lodewikus Lee)', dated 24 April 2000 (Blaauwberg Tourism Office); a brochure entitled 'Discover the Battle of Blaauwberg', compiled by Pat Matejek (2002); 'The Blaauwberg Battlefield Revisited', *Cape Times*, 3 April 1926; a typescript memorandum from Mansell Upham to Pat Matejek; a typescript school project on the restoration of Rietvlei Spring by Inge van Egeren et al.; a pamphlet entitled 'Description of the Battle of Blaauwberg' (Blaauwberg Tourism Office).

Chapter Six: Slagtersnek
The primary source is H.C.V. Leibbrandt, *The Rebellion of 1815, generally known as Slachters Nek* (Cape Town, 1902; a government publication with the classification number A465). This is a complete collection of all the papers connected with the trial, with many important annexures. Indispensable is J.A. Heese, *Slagtersnek en sy Mense* (Johannesburg, 1973). Also consulted were J. Albert Coetzee, *Skavot* (Pretoria, n.d.) and J. Albert Coetzee, *Ruiters van Slagtersnek* (Pretoria, 1949). Also important is the five-page unpublished manuscript of Charles Harte (with an additional annexure by Gwyneth Harte) preserved in the Somerset East museum.

Chapter Seven: Amalinde
Aside from Mostert (pp. 442–491) and Milton (pp. 68–69), a key text is J.H. Soga, *The South-Eastern Bantu* (Johannesburg, 1930), particularly chapter 13, pp. 148–177. A.K. Soga's poem is taken from T. Couzens and E. Patel (eds.), *The Return of the Amasi Bird: Black South African Poetry 1891–1981* (Johannesburg, 1982), p.17.

Chapter Eight: Grahamstown
See Cory, v. 1, chapter 12, pp. 369–403; Mostert, chapter 13, pp. 442–491; Milton, chapter 9, pp. 65–75. See also a pamphlet issued by Grahamstown Tourism entitled 'Egazini, Place of Blood: The Battle of Grahamstown, 22 April 1819'; and David Owen, *Ubukhosi neenkokeli* (Grahamstown, 1994).

Chapter Nine: Lattakoo
An important source is I. Schapera (ed.), *Journal and Letters of Robert and Mary Moffat* (1951), pp. 86–101. See also W.C. Watson, 'Takoon 1823: An Exercise in Map Collection' in *The South African Survey Journal*, v. 13, no. 78, December 1971, pp. 28–35; W.F. Lye (ed.), *Andrew Smith's Journal 1834–1836* (Cape Town, 1975), pp. 194–197. Some information came from a brochure entitled 'Kuruman: The Oasis of the Kalahari', produced by the Ga-Segonyana Municipality. Also consulted was P.H.R. Snyman, *Olifantshoek* (Pretoria, 1956).

Chapter Ten: Houdenbek
On the slave revolt see Dene Smuts and Paul Alberts, *Die Vergete Grootpad* (Bramley North, 1988).

Chapter Eleven: Khunwana
On the battle of Khunwana itself, see S.M. Molema, *Montshiwa: Barolong Chief and Patriot, 1814–1896* (Cape Town, 1966), pp. 11–23. For the novel, see Sol T. Plaatje, *Mhudi* (Cape Town, 1996). For a fine biography of Plaatje, see B. Willan, *Sol Plaatje: A Biography* (Johannesburg, 1984). For other information and interpretation of the battle see T. Couzens, introduction to *Mhudi* (Heinemann, London, 1978) and T. Couzens, 'Editors' Views' and additional material in *Mhudi* (Francolin Press, Sefika Series, Cape Town, 1996), pp. 159–189.

Chapter Twelve: Salem
See Mostert, p. 673; Milton, pp. 103–109; F.C. Metrowich, *Assegai over the Hills* (Cape Town, 1953), pp.68–78. There is also a useful pamphlet entitled 'Salem: A Walkabout' which can be purchased in the Salem church. For general background see, for instance, Guy Butler (ed.), *The 1820 Settlers: An Illustrated Commentary* (Cape Town, 1974).

Chapter Thirteen: Congella
There is an extremely useful unpublished typescript entitled 'Diary and Notes connected with the 2nd British Military Occupation of Port Natal 1842' by Stanley Evans in the War Museum Library in Johannesburg (Classification number Q968.04EVA). Unfortunately it is incomplete and I have not been able to track down the author. Also useful: 'An Exposition of the Clash of Anglo-Voortrekker Interests at Port Natal leading to the Military Conflict of 23–24 May 1842', by A.E. Cubbin in *Historia*, v. 37, no. 2, November 1992, pp. 48–69; J.D. (full name unknown), 'The Old Fort at Durban' (Durban, 1910, published for the South African National Society); T.V. Bulpin, *Natal and the Zulu Country* (Cape Town, 1966), pp. 93–137; Ian Knight, 'The Siege of Durban, 1842' (provenance unknown, copy in War Museum Library, Johannesburg); Theal (1908), v. 2, pp. 346–378; and Cory, v. 4, pp. 102–207. H.I.E. Dhlomo's poem is taken from N. Visser and T. Couzens (eds), *H.I.E. Dhlomo: Collected Works* (Johannesburg, 1985), p. 363. There is a photograph of Ndongeni in later life in Cory (opposite p. 154). A precis of Ndongeni's account of the ride appears on p. 155.

Chapter Fourteen: Boomplaats
Amongst the sources consulted were: *The Autobiography of Lieutenant-General Sir Harry Smith*, edited by G.C. Moore Smith (London, 1902), v. 2, pp. 238–250; Major G. Tylden, 'Boomplaats, 29th August, 1848', in *Journal of the Society for Army Historical Research*, v. 16, pp. 207–213; J. Falkner, 'Battle at Boomplaats', *Military Illustrated*, no. 100, September 1996, pp. 54–55; J. Loock and P. Kennedy, 'Die Slag van Boomplaas', pamphlet, 1998; Theal, v. 3, chapter 45, pp. 260–285.

Chapter Fifteen: Fort Beaufort
Most detailed here is N. Mapham, 'The Battle of Fort Beaufort', *Martello newsletter*, no. 25, March 1980, pp. 2–7; for the attacks on the farms, see 'The Attack on the Hammonds', in the *Martello newsletter*, no. 41, June 1984, pp. 9–16; and for information on Sipton Manor see *Martello newsletter*, no. 45, September 1985, pp. 3–13. Milton also has a short section on the battle, pp. 192–194.

Chapter Sixteen: Centane Hill
Mostert is outstanding on the story of Nongqawuse. Information was also gleaned from A.W. Bruton, *Sparks from the Border Trail* (King William's Town, n.d.). Milton has a key description of the battle at pp. 267–269.

Chapter Seventeen: Thaba-Moorosi

A copy of the letter by the 'unknown soldier' (a 30-page manuscript) was given to me by David Ambrose of the National University of Lesotho. Other critical information comes from the book *Twenty-Five Years' Soldiering in South Africa* written by 'A Colonial Officer' (London, 1909) (it is probable that this was H.V. Woon); also G. Tylden, *The Rise of the Basuto* (Cape Town, 1950), pp. 128–144.

Chapter Eighteen: Kanoneiland

The principal source for this chapter is Teresa Strauss, *War along the Orange: The Korana and the Northern Border Wars of 1868–9 and 1878–9* (Centre for African Studies, University of Cape Town, 1979). It deserves wider circulation. Another important work is J.A. Engelbrecht, *The Korana* (Cape Town, 1936). See also, W.A. Burger, *Kenhardt uit ons Geskiedenis* (unpublished typescript in Upington Museum).

Chapter Nineteen: Potchefstroom

The key participant account, at least from the British side, was R. Winsloe, 'The Siege of Potchefstroom' (pamphlet, no place, no date), 39 pp. This needed to be supplemented by Ian Bennett's excellent book, *A Rain of Lead* (London, 2001) and the useful summary of the siege by Gert van den Bergh in his *24 Battles and Battle Fields of the North-West Province* (Potchefstroom, 1996), pp. 28–40. It is deeply to be regretted that the latter's full account of the siege has not yet been published. Useful information was also gleaned from the pamphlet, 'Chevalier Oscar Wilhelm Alric Forssman 1822–1889' (Pretoria, 1962) by Alric Forssman; 'Magnus Johan Frederik Forssman 1820–1874', (no place, no date) by Alric Forssman; W. Prinsloo, *Potchefstroom 150* (Potchefstroom, 1989); and Elsa Smithers, *March Hare* (London, 1935). An important study, too, is Julian Orford, *95 Days* (Potchefstroom, 1973). There is also a pamphlet on the Goetz-Fleischack Museum provided by the Potchefstroom Museum.

Chapter Twenty: Blouberg

This chapter is centred on Revd. Colin Rae's diary-based account, *Malaboch* (London, 1898). Also important: J. van Schalkwyk and S. Moifatswane, 'The Siege of Leboho: South African Republic Fortifications in the Blouberg, Northern Transvaal', in *Military History Journal*, v. 8, no. 5, 1991, pp. 175–183. See also, Noel Roberts, 'The Witchdoctor's Drums', in *The Star*, 20 August 1932 and Anon., 'War Drums of Malaboch' (newspaper article, provenance unknown, probably *The Star*), as well as two letters of P.W. Cahill in Museum Africa (Add. Note 46/354 and Add. Note 46/202).

Chapter Twenty-One: Magoebaskloof

The best description of the life and death of Makgoba is Louis Changuion's booklet, *Makgoba (Magoeba) and the War of 1895* (1999, privately printed). See also L. Changuion, *Haenertsburg 100* (Pietersburg, 1987) and Harry Klein, *Valley of the Mists* (Cape Town, 1972). For a fuller discussion of Buchan's novel, see

T.J. Couzens, '"The Old Africa of a boy's dream": Towards Interpreting Buchan's Prester John', in *English Studies in Africa*, v. 24, No. 1, 1981, pp. 1–24.

Chapter Twenty-Two: Thaba-Bosiu
For a large range of sources and a background to the long history of Thaba-Bosiu see Tim Couzens, *Murder at Morija* (Johannesburg, 2003). A specific debt is owed to G. Tylden, *History of Thaba-Bosiu* (Maseru, 1945), as well as to David Ambrose, 'Masupha's Village', in *Lesotho: Basutoland Notes and Records*, v. 9 (1970–71), pp. 20–24. Also consulted was F. Wepener, *Louw Wepener* (Pretoria, 1934).

Chapter Twenty-Three: Tafelkop
For details of the battle, see *The Times History of the Anglo-Boer War*. T. Mofolo's *Moeti oa Bochabelo* was published in book form in Morija in 1907. For a study of Mofolo see D.P. Kunene, *Thomas Mofolo and the Emergence of Written Sesotho Prose* (Johannesburg, 1989). See also, Tim Couzens, *Murder at Morija* (Johannesburg, 2003), particularly chapters 21 and 23. Information concerning the early history of the Basotho and Ntsoanatsatsi derives in part from D.F. Ellenberger, *History of the Basotho* (Morija, 1992).

Chapter Twenty-Four: Nakob
Inspiration for this chapter came from F. Cornell's fascinating book *The Glamour of Prospecting* (Cape Town, 1986), edited by D. Cornell, introduction by T. Couzens. See also, T. Dedering, 'The Ferreira Raid 1906: Boers, Britons and Germans in southern Africa in the aftermath of the South African War', Unisa Library Conference, 3–5 August 1998; J.J. Collyer, *The Campaign in German South-West Africa 1914–1915* (Pretoria, 1937); 'The Empire's War for Freedom', in *South Africa, 31 October, 1914* and *The Union of South Africa and the Great War 1914–1918 Official History* (Pretoria, 1924).

Chapter Twenty-Five: Keimoes, Kakamas and Upington
Crucial documents are the *Report of the Judicial Commission of Inquiry into the Causes and Circumstances relating to the Recent Rebellion in South Africa* (Cape Town, 1916) U.G. 46–'16; and *Report of the Outbreak of the Rebellion and the Policy of the Government with regard to its Suppression* (Pretoria, 1915) U.G. 10–15. Information about the battle at Keimoes also comes from Maria de Beer, *Keimoes en Omgewing* (Keimoes, 1992), pp. 107–111. On Kakamas see H.C. Hopkins, *Kakamas – Uit die Wildernis 'n Lushof* (Cape Town, 1978), pp. 128–129; and assorted documents in the town library. *Agter die Skerms met die Rebelle* by 'Kalahari Mac' was published in Johannesburg in 1949. Also useful is A.K. Cornelissen, *Langs Grootrivier* (typescript).

Chapter Twenty-Six: Potter's Hill
The bulk of the material for this chapter came from contemporary newspaper reports (mainly *The Star* and the *Rand Daily Mail*).

THANKS

This book is dedicated to Charles van Onselen. Perhaps it will be my last, so, Charles, it must be yours. Others who have had a continuous influence on my work, past and present, are Tony Traill, Marshall Walker, Belinda Bozzoli, Ivan Vladislaviç and Brian Willan. Those who have had an effect on this book, direct or indirect, are Michelle Adler and David Bunn (my colleagues in a travel writing course – they are terrific teachers), Gerda and John Wollheim, Nigel Fox, James Clarke, Karen Botha, Kate and Ken Owen, Jenny Hobbs, James Christie, Hugh Cecil, Hal and Edmund Couzens and Per Bohlin.

This kind of research is extremely expensive and I am grateful to the National Arts Council of South Africa for a Writer's Grant which enabled me to complete this work. I am grateful, too, for the support of my publishers and of the Graduate School of Humanities and Social Sciences at the University of the Witwatersrand (the director, Carolyn Hamilton, provided me with an intellectual home over the last three years while I researched and wrote *Battles*).

I am particularly grateful to Nicholas Combrink and Basil van Rooyen who initiated the project and to Sharon Hughes and Jeremy Borraine who have offered continuous encouragement. Pat King and my wife, Diana Wall, have shouldered the logistical production of the draft of this work way beyond what I deserve. And Angela Briggs (editor), Karen van Eden (project manager), Ann Westoby (maps), Tessa Kennedy (proofreading) and Val Myburgh (drawings) put the book to bed with quiet efficiency and infinite patience. Bless them all.

Bill Nasson, fine academic as he is, read the manuscript with a critical, but humane, eye and made many helpful suggestions. Some of them I could not incorporate for practical reasons, so some of the deficiencies in the book have to be swallowed in spite of his advice to the contrary. He is not to blame for anything that is here or not here.

I would like to thank, too, an unusual group of people whom I have never met but who have been my constant companions and touchstones even in the darkest days and undoubtedly will continue to be so into the darkest night: Geoffrey Chaucer, George Orwell, Joseph Heller, James Joyce, Solomon Plaatje, Lawrence Sterne and Herman Melville. They have more integrity, each of them individually, than all of the politicians, combined, who ever lived and lied (this, of course, being a tautology).

Amongst the general thanks, finally, I would like to express my particular gratitude to my daughter Kate Couzens Bohlin who has encouraged me endlessly in this endeavour. Thank you, Kate. And thank you, Diana, my beloved wife, for helping me realise that a lot of short trips are not as important as the long journey.

*

Amongst the scores of people who have helped me in my research for *Battles* I would like to record my thanks to the following:

Gerrit Olivier (for helping me with the translation from Dutch of De Kuiper's journal); Ross Parry-Davies (Hout Bay); Cecil Ford (Posthuys, Muizenberg); Pat Matejek, Pete and Megan Reinders (Blaauwberg); Lieutenant-Colonel Ian van der Waag, Marius Meiring and Chris Mathys (Saldhana Bay); Sheila van Aardt, Kevin McCaughey, Hannes Nieuwoudt, and Bill and Allison Brown (Somerset East); Marco and Olenka Brutsch (Grahamstown) and Tshawe Nyingwashe and Cikiziwa Sauli (Debe Nek); Di Tremeer (Alice); Professor Colin Coetzee (Grahamstown); Julie Wells (Grahamstown); Richard Aitken and Jane Argall (Kuruman and Dithakong); Pieter van Wijk (of the farm Gamasip, Langsberg); Maureen and the late Brian Bamford (Koue Bokkeveld); Dene Smuts; Carel van der Merwe (Boplaas farm); Chilly Ramagaga (Moroka's Hoek); Brian and Rose Long (Salem); Charles van der Merwe (Old Fort Museum, Durban); Jane and Jon Roberts (Baddaford Farm), Gert and Mattie van der Westhuizen (Olive Cliff Farm) and Shirley and John Sparks (Leeuwfontein Farm); Peggy Veldman (Shiloh Mission, Whittlesea); Dave Dauncey (Fort Cox); Trevor Wigley (Trevor's Trails and Adventures, Qolora Mouth) and Rod Dewberry (Seagulls Hotel); David Ambrose, Stephen Gill and Albert Brutsch (Lesotho); Hennie Steyn (Kanoneiland); Gert van den Bergh (Potchefstroom); Louis Changuion (Haenertsburg); Patrick McGaffin (Magoebaskloof); Bertie de Jager (Tafelkop); Willem Naudé (Vrede); the late Doug Cornell (son of Fred Cornell); Jopie Kotze (Springbok); Charlotte Viviers (Memel); Quentin Coaker; the late T.V. Bulpin; Cynthia Kemp; James Mitchell (*The Star*); Mike Birch (of Double Drift Game Reserve, who showed me Fort Willshire when I could never have got there on my own).

I owe a great deal to numerous libraries and museums and their overworked, understaffed and underpaid staffs: in particular, the library of the War Museum in Johannesburg; Museum Africa (Johannesburg); Saldhana Bay Public Library; the South African National Library; Kakamas Municipal Library; Upington Municipal

Library and the Upington Museum; the Old Fort Museum, Durban; the University of Cape Town manuscript collections (Fred Cornell papers); the Albany Museum and the Observatory Museum (Grahamstown).

I'd also like to single out individuals in this respect: Emile Badenhorst (Somerset East Museum); Moose van Rensburg (Fort Beaufort Museum); Zwelinyanyikima Vena (Cory Library, Rhodes University); Stephanie Pienaar (Amatola Museum, King William's Town); Christine Driewes (Potchefstroom Museum); Poekie McSporran (Castle Road Museum, Port Elizabeth); Linda Swart (Pietersburg Museum); Margaret Noth (University of the Witwatersrand William Cullen Library); Jurgen Witt (Tzaneen Museum), and Diana Wall (Museum Africa).

The owners of Fuller Farm, Ghalla Hills and Boomplaats allowed me with gracious hospitality to gatecrash their farms. Stephen Fick of Milkwood Dairy, Alexandria, kindly showed me Nongqawuse's grave.

A number of tourism offices also provided valuable information: these include Upington, Kenton (Erica McNulty), Kuruman, Somerset East, Port Elizabeth, Grahamstown and Pietersburg. Star of the lot is Pat Gee of the tourism office at Table View (and Bloubergstrand). She is my ideal of a tourism officer. Places like King William's Town have a lot to learn from her about how to handle visitors in a place that has a lot to offer.

I have also benefited from the help and support over the last few years of various booksellers, publishers and journalists and reviewers. They will know who they are and I thank them.

Finally, five of these chapters (numbers 10, 11, 21, 23 and 26) first appeared as travel articles in the *Lifestyle* supplement of *The Sunday Times*. I would like to thank it and its editor, Stephen Haw, for giving me the chance to experiment with what was for me a new genre.

Obituary

As this book goes to the printers I learn of the sad death by fire of the Magoebaskloof Hotel. Readers will be pleased to hear that Chief Makgoba's head has not been lost for a third time. For 30 years the hotel has been one of my favourite places on earth (see Chapter 21). I hope it will rise again, literally from the ashes, not as some modern monstrosity, but with its old charm.

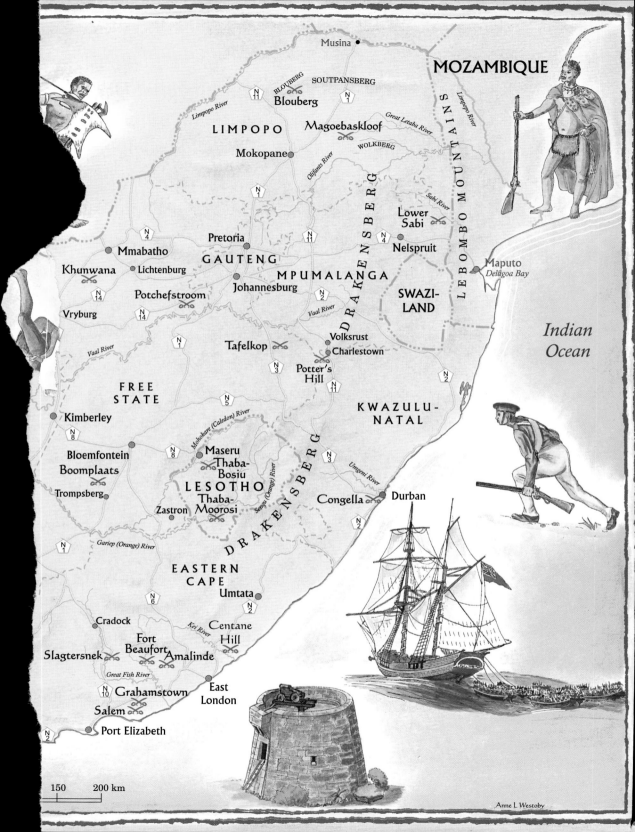

Musina

MOZAMBIQUE

BLOUBERG
SOUTPANSBERG
Blouberg

LIMPOPO

Magoebaskloof

Great Letaba River

Mokopane

WOLKBERG

Olifants River

Sabi River

Lower
Sabi

LEBOMBO MOUNTAINS

Limpopo River

N
4

Pretoria

GAUTENG

Nelspruit

Maputo
Delagoa Bay

Mmabatho

Lichtenburg

MPUMALANGA

SWAZI-
LAND

Khunwana

Potchefstroom

Johannesburg

Indian
Ocean

Vryburg

N
14

N
14

Vaal River

Vaal River

N
2

Volksrust

Tafelkop

Charlestown

N
1

N
3

Potter's
Hill

N
11

N
5

KWAZULU-
NATAL

Kimberley

FREE
STATE

Mohokare (Caledon) River

N
8

N
8

Maseru
Thaba-
Bosiu

N
2

Bloemfontein

Boomplaats

LESOTHO

Thaba-
Moorosi

N
3

Umgeni River

Trompsberg

Zastron

Congella

Durban

N
2

Gariep (Orange River)

Senqu (Orange) River

DRAKENSBERG

N
1

EASTERN
CAPE

Umtata

Cradock

N
6

Centane
Hill

N
2

Kei River

Fort
Beaufort

Amalinde

Slagtersnek

Great Fish River

East
London

N
10

Grahamstown

Salem

Port Elizabeth

N
2

DRAKENSBERG

150   200 km

Anne L Westoby

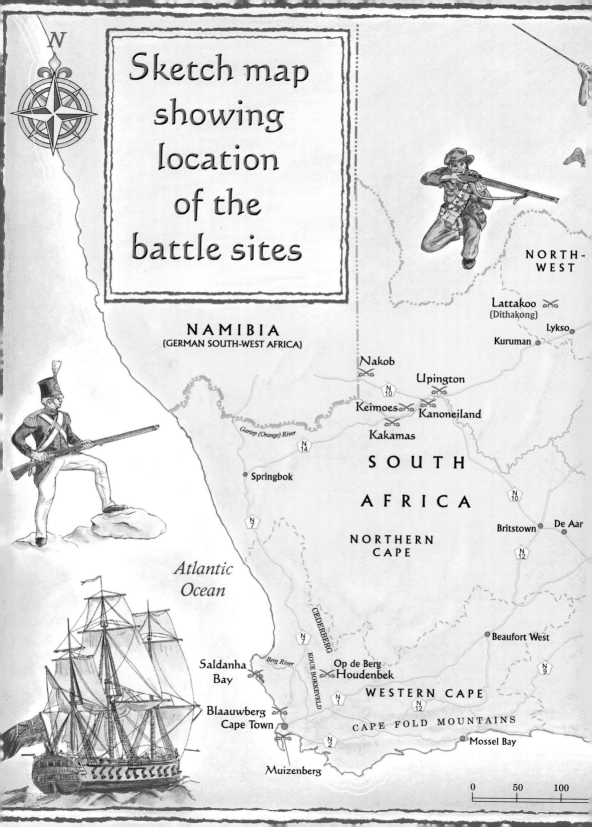

Sketch map showing location of the battle sites

N

NORTH-WEST

Lattakoo
(Dithakong) ⚔

Lykso

Kuruman

NAMIBIA
(GERMAN SOUTH-WEST AFRICA)

Nakob ⚔

Upington ⚔

N
10

Keimoes ⚔

Kanoneiland

Kakamas

Gariep (Orange) River

N
14

SOUTH

AFRICA

N
10

Springbok

N
7

De Aar

Britstown

**NORTHERN
CAPE**

N
12

*Atlantic
Ocean*

Beaufort West

CEDERBERG

N
7

Berg River

Op de Berg ⚔
Houdenbek

N
9

KOUE BOKKEVELD

Saldanha
Bay ⚔

N
1

WESTERN CAPE

N
12

Blaauwberg ⚔
Cape Town

CAPE FOLD MOUNTAINS

N
2

Mossel Bay

Muizenberg

0   50   100